BUILDING WITH HEART

CHRISTOPHER DAY trained in architecture and sculpture. In addition to designing buildings, he has built seven of them, including two Steiner schools with volunteer groups. Influenced by Rudolf Steiner to develop a fresh way of looking at situations with wide and practical applicability, he is in international demand as a lecturer. He is also the author of *Places of the Soul* (Thorsons Publishing, 1990).

The author at Nant-y-Cwm Steiner School.

BUILDING WITH HEART

A PRACTICAL APPROACH
TO SELF AND COMMUNITY BUILDING

CHRISTOPHER DAY

GREEN BOOKS

First published in 1990 by
Green Books
Ford House, Hartland
Bideford, Devon EX39 6EE

Typeset by Computype, Exeter.

Printed by Hartnolls,
Victoria Square, Bodmin, Cornwall.

Printed on 100% recycled paper.

British Library Cataloguing in Publication Data
Day, Christopher
Building with heart : a practical approach to
self and community building.
1. Great Britain. Self-build housing
I. Title
363.58

ISBN 1-870098-03-X

Contents

Acknowledgements

I would like to thank Caroline Kedward for typing, and endlessly re-typing, the manuscript; David Hockley, Vicky Moller and Rosemarie Maspero for reading and commenting on the draft; Sabine Roberts, Nim de Bryne, Julian Bishop, Paul McIntyre, William Browne and Ulrike Hotz for photographs; and also Julian Bishop, David Hockley, Gareth Roberts and Nick Dominney and William Browne for assistance with the design and/or execution of some of the projects mentioned.

And last but not least, I am most grateful to the many volunteers with whom, for long and short periods, I have worked.

A Note from the Author

My grandmother used to say 'Do as I say, don't do as I do', and I am only now coming to realise the excellence of this advice. This book is about the vision to which I aspire. It is well grounded in practical experience; successful and not so successful.

Brooding upon my own weaknesses and failures, with regard to the ideal, would obscure the possibility for good which I believe is inherent in this vision. I believe that neither the vision nor the book is dishonest – but the reader must understand that these are ideals for which I strive. That I do not always live up to them, goes without saying.

My grandmother also used to say that it was impossible to twiddle your thumbs in both directions as once. Some people see this challenging riddle as a solution to unemployment – but I have given up trying.

Preface: making the impracticable practical

I was trained in architectural school to try to like modern architecture – then at its functionalist, rectilinearist peak. Over the years, I became increasingly horrified by the results of such architecture as I began to realize what a harmful influence it had upon people and upon society. I began to wonder: if architecture can be so powerful and destructive, could it not also be beneficial and healing? Healing is not dominating. Healing and dominating power are opposite ways of going about things; healing allows the individual to create. Power, from the outside, destroys.

Over twenty years ago I became aware that architecture is rarely experienced in isolation. Awareness of the dialogue between place and building is an important concern, which I try continually to refine. Nowadays I try to listen also to that which is not directly seen or heard, that which lies behind the world accessible to the outer senses. In people and society, as in places, it is this invisible spirit which is at their essence. Listening is the first step in any therapeutic process.

Through the experience of building my own houses, I was able to allow myself to shed my functionalist prejudices and build as I had formerly made sculpture – buildings that extended sculpture to become whole environments.

With this expertise, it was natural to extend building by and for ourselves into building with other volunteers for others when I became involved in the founding of a Steiner school. The new elements of gift and community brought to the fore issues which were previously just in the everyday stream of things. It is out of this experience, and the necessity of developing it consciously, that this book has been written. Here, art, healing, gift, and listening are combined as a social work within the work of building.

Through my building experience with other volunteers those values that I have carried over the years have met with insights which were brought by the situation.

I write therefore of gift-work building and what it means to arrange and organize it. In doing so I am writing of how building can be released from conventional constraints to allow itself to be an artistic work, to unveil its healing potential.

Some practically inclined people feel that only the solid practical material is important; the airy-fairy philosophizing can be left out. I used to feel like that, but no longer. Solid practical work has caused me to revise my attitude. Indeed, to work meaningfully and relevantly, to work with gift, we have to reverse the everyday order of priorities, to stand conventional attitudes and often processes on their heads.

In Love lives the seed of Truth
in Truth seek the root of Love:
Thus speaks thy higher Self.

The fire's glow transmutes
Wood into warming rays.
Wisdom's resolving Will
Changes the outer work
Into abiding strength.

So let thy work be the shadow
Cast by thine I
When it is lit by the flame –
Flame of thy higher Self.

Rudolf Steiner, *Verses and Meditations,* Rudolf Steiner Press, London, 1972, p.119.

You have been told also that life is darkness, and in your weariness you echo what was said by the weary.
And I say that life is indeed darkness save when there is urge,
and all urge is blind save when there is knowledge,
and all knowledge is vain save when there is work,
and all work is empty save when there is love;
And when you work with love you bind yourself to yourself, and to one another, and to God.

And what is to work with Love?
It is to weave the cloth with threads drawn from your heart, even as if your beloved were to wear that cloth.
It is to build a house with affection, even as if your beloved were to dwell in that house.
It is to sow seeds with tenderness and reap the harvest with joy, even as if your beloved were to eat the fruit.
It is to charge all things you fashion with a breath of your own spirit,
and to know that all the blessed dead are standing about you and watching.

Kahlil Gibran, The Prophet, Alfred A. Knopf, New York, 1963, pp.26.27.

I

A STARTING POINT

Ty-cwrdd Bach – the first house I built. As it was as a ruin before we started work and as it is now. It was my first lesson that building is part of the artistic process. On site I realized that I could create a curved roof by setting rafters between non-parallel wall-plates. The curve that resulted I could never previously have imagined on paper. Only on site could I see the potential to develop it.

1

Towards Volunteer Building: a personal journey

IT WAS in 1972 that I gave up my job teaching architecture in London and returned to Wales to build a house for my wife and myself. After months of searching we had found a ruin in an area and community I know well from my childhood. It really was a *ruin*. Once a chapel, it now comprised three three-quarter walls and one quarter wall with nine small trees and a mass of bramble growing within its shell. Stones fallen off the walls formed an overgrown mound around their base. The ruin had grown into the landscape.

This relationship with the landscape struck an immediate chord in me. This was how buildings should be in the landscape – timelessly belonging. Would it be possible to build a home out of this ruin and yet retain its essential qualities – quietness, unobtrusiveness, the feeling of being rooted in the earth, above all this quality of timeless belonging to its surroundings?

With a naïve enthusiasm I approached the task of building.

I brought with me a knowledge of textbook building

1

construction and the boyhood experience of building sheds on my father's smallholding. I assumed this background would be adequate. It was not.

That it was not, was my good fortune. The limitations of my background proved, in fact, to be an invaluable asset. Not knowing the range of standard items that could be bought off the shelf, nor the proper standard way of doing things, I had to work out how to make things, how to do them. This process took time and concentration. Time and concentration gave the opportunity and the attunement to examine, select and refine the aesthetic alternatives.

With the help of friends and the appreciative encouragement of neighbours, my wife and I slowly created the building.

As it advanced, so its slow rate of progress and the fact that 'how to' achieve a function was always in mind, meant that quite naturally the design evolved. Design, constructional principles and the unfamiliarity of the work remained alive throughout the whole project. At no time were they stultified by fixed plans, catalogue components or routine practice.

Without being aware of it, I was beginning to engage head and heart as well as hands in what I thought of as purely a physical exercise. It was also my first introduction to the fact that building itself can be an art.

Art as a process is a chalice of opportunity. No painting of worth can be executed exactly to a predetermined plan. It develops a life of its own. Take up the opportunities and enhance them and the painting grows in inner value. Obliterate them to conform with a plan and its value shrinks. So also with building, I discovered. Six years later I was to realize that the working relationships on a volunteer site lend themselves to just such a process.

A shortcoming of my background was that I had no experience in organizing work – my own or anyone else's. This led to much needless effort and inefficiency. All the mortar and even much of the concrete was carried in buckets across uneven and bumpy ground, stone piles and the like. Stones and concrete blocks were carried by hand. Over the whole job the quantity must have been in the region of 100 tons. Nowadays I would prefer to spend a week moving the stones sufficiently away from the building – even though they would now be two yards (three laden paces) from where they would be wanted – so that everything could travel around the site by barrow.

It was a hard way to learn a basic lesson.

It also increased the heavy drudgery component of the work, already considerable due to our desire not to use machinery (other than a JCB to dig the septic tank hole). In those days I regarded work as labour – transformation, I thought, only comes at the end – and I expected friends who came to help us to labour, as I did myself.

Many friends came to visit us – and all of them either expected or were expected by myself, to work. Without their help I doubt if I could ever have sustained the will necessary to accomplish the task. Some we knew were coming. Most just turned up. We had no telephone, and our accommodation consisted of a chicken shed about 9 feet by 5 feet in floor area, a similar shed for tools and materials and a tent for visitors. It was a windswept site and the weather often chose to slash with rain when visitors arrived. There was no sheltered work so we would retire to the chicken shed in bad weather. And when it didn't rain we worked. While I have never become sufficiently efficient at it, it was a good preparation for working with unpredictable numbers of volunteers.

The work became a race against the bad weather. Could we get a roof on – to protect the roof carpentry, more even than ourselves – before the autumn gales became the winter rains? Typically we worked every daylight hour, seven days a week, without even the ease, smoothness or effectiveness which comes with familiarity with the work. Building began to feel an endless burden.

Occasionally friends took photographs – we were too busy to. The photographs only emphasized what I had already become aware of. Our small achievements each day as blockwork began to take shape within the stonework were *so ugly*. At the end of each day, what could I look back on that was a benefit to anyone other than ourselves, and what of beauty?

That first summer and autumn was an exhausting period. The endless hard labour was not enlivened by rhythm or redeemed by beauty. There was no nourishment. I have since come to realize that work without spiritual nourishment, without beauty, strains

3

the will to the utmost. In the conventional world, the will is bribed by material gain: pay, profit or possession. Volunteers work to fulfil a pressing need. But always, without beauty, the work is harder, heavier, duller – a drain on one's energies.

Next followed a dreary and ineffective period in the rain and gale-lashed winter gloom. At this time we had only paraffin lights and candles. Not until the next year did we get a wind generator to light the house.

At last we could plaster the blockwork and make windows and doors. I had never made a window before, but as I couldn't buy one that wasn't rectangular, I had to learn. It wasn't so hard. Internal doors, not having to stand up to weather were, I discovered, easy. Now, with the finished stonework, unifying paint, and shelves and cupboards on the way, the work at last began to feel artistic. No longer drudgery, it was a joy to bring such transformation into being.

At the same time I began to receive architectural commissions. We had started building on my small savings, followed shortly by half the proceeds of the sale of my grandmother's house (which I had half inherited). Now I began to finance the building out of my work. By the end of the job I had begun to learn some lessons that should have been obvious from the outset. I had come to realize that all the many friends who helped, gave their work as a gift; and that I had a responsibility to repay them by helping their work become a fulfilling experience. I had learned that if work is not a pleasure, it is a drain on one's energies; that it cannot be a pleasure if its effects are unrelieved ugliness. Only when the result of a day's work was attractive did the work change from drudgery to fulfilment. I had begun to experience the freedom, the enjoyment and the aesthetic necessity of making, rather than buying, a wide range of components from window arches to door latches, ventilators to sink units.

Ty-cwrdd Bach was originally a meeting-house. Echoes of its former function endured. It was a house of many meetings, including those chance meetings which subsequently led to the foundation of Nant-y-Cwm Steiner School. I had approached the work very heavily – as a material task. The spirit of the house rescued me from being submerged by the heavy, wholly

4

unmechanized labour of long hours of unfamiliar, ineffectively organized work.

It is a good house and good experience to look back on. But it was hard work.

I swore it was the last house I would ever build.

* * *

DURING 1975 we began to realize that it is impossible to keep animals without land. My wife would take our goats long distances to the woods to graze. On one occasion they saved us the effort of going to collect them, by making their own way home. The next day we found out that the boredom of their journey had been relieved by dancing on every roof in the valley. We heard the story from a neighbour who had been entertaining the vicar to tea at the time. Suddenly there was a noise overhead like the devil drumming on the slates. They rushed outside and caught a glimpse of cloven hooves . . . .

Either we had to give up animals or move to a house which had land. It was this that led us to Pen-y-Llyn.

Pen-y-Llyn was more orthodox as a ruin. It looked like a farmhouse but was in fact so deteriorated that little more than three walls could be retained and not even all of them were sound enough not to need rebuilding.

Its last occupants had been caravan dwellers who fell out with their landlord, vandalized the house and mutilated its trees. In the sixteen years since it last was a home a dark spirit had come to lie heavily on the place. Our work was principally work to redeem it.

With the benefit of experience I arranged access for cart and barrow and space for scaffold, located sand, gravel, block, brick and stone heaps for maximum convenience, bought a hand-powered mixer and employed a friend to work as my mate. Although I was the more skilled, he and I worked as social equals. The hierarchical concept of craftsman and labourer seemed out of place and I began to evolve techniques of jointly working on both craft-work and labouring. These techniques became the foundation of the teamwork techniques so essential to future volunteer building. I was very lucky that circumstances had led me along that path.

Initially work went very smoothly, but as time went on, unforeseen problems, particularly with the existing structure, began to appear. My assistant suffered from ill health and I was frequently on my own at times when tasks required a team of two. I was financing the job with my architectural work, effectively a 36-hour week, which had to be fitted into weekends, Mondays and before breakfast and after supper, every day. I was also under intense pressure to complete within a year due to house-improvement grant requirements. Although grants are administered by the Environmental Health Department, for many self-builders the pressures imposed by grant time limits strain physical, mental and family health to or even beyond their limits. At this stage we had two young children.

As more and more things went wrong, I began to sense that the dark spirit did not want to leave the place.

For most of the large concreting jobs I hired a mechanical mixer and for the drainage and piping to our water-powered generator I hired an excavator. It broke down so many times that the driver got fed up with the job and never returned to fill in the trenches. Most of these we had to do by hand and my body has never quite recovered. At the very end of the job – too late to use for building – I even acquired a tractor and link box. Had I had this earlier I could have cut the labour of moving materials around the site to a quarter.

The work was hard and long, but with sufficient experience now to be able to work smoothly, rhythmically, productively and attractively, I had begun to enjoy it. The building, the surroundings and the land began to be redeemed.

But it *was* enough work! I was never going to build anything again!

But by the time we moved into Pen-y-Llyn, we had met a number of other families who wanted a Steiner education for their children. Circumstances set in motion the founding of Nant-y-Cwm Steiner School.

As a group we purchased an old Victorian school building, which had been closed for twenty years and was suffering from long neglect. To repair, renovate, provide essential services and extend its volume was a massive job, and because we had no money could only be undertaken by volunteers. The experience I had gained in six years of building and in my architectural practice, in which I was frequently designing for and advising self-builders, became a foundation for working with volunteers. This was a field of which I knew nothing, blithely assuming that if it was possible, we would do it. We did do it, and it proved both harder and easier than I could have imagined.

It was harder because of the volume of work involved and because I had gravely underestimated the 'promise factor' – the proportion of promises to help which is actually realized in sustained effort. I had assumed it would be around 5:1. A fourfold underestimation, but it is difficult to give the promise factor a fixed value. In some circumstances and with some individuals promises are 100 per cent fulfilled, in other cases less than 5 per cent. I now know what and when to ask, and whom I can rely on. As we started work, two of us, in the winter rains and mud, with no money to buy materials, we soon became joined by others, and supported by donations often from complete strangers. We learnt that unfulfilled ideas produce enthusiasm that may be unsustainable but commitment in deed inspires committed effort.

I once thought that if enough fish swam together, they could turn the current. There were never enough fish. There were never enough money, enough volunteer builders, enough time – but the initiative wanted to come into being. Our deed enabled a spiritual reality to take root in the material world. Now I know that the current is *waiting for the opportunity* to turn. I have seen it proved at Nant-y-Cwm and I can see it in so many spheres in the world around us today.

Much that I had never previously been aware of, I learnt at Nant-y-Cwm. I learned that volunteer building enables things to be done that are virtually impossible within a contract system. I learnt that not only can buildings be built cheaply enough to be affordable by charities with little money, but that it costs no more if the time is taken to imbue the building with a truly artistic quality. Indeed I learnt that it would be hard to sustain the effort over the long time span required were not the work itself in some way an artistic experience. Volunteer work lends itself to this in that it is neither bound by the costs of labour nor by the inflexibility of a contract.

As the building rises and becomes formed, so many potentials, invisible on the drawing-board, unfold. Design, freed from the constraints of a paper monologue, can become a process of conversation. This conversation can unlock the potential inherent in the place, the building, and the people involved. It can allow the craftsman to blossom into the artist, re-establishing the tradition of architecture as art* and serving the human being. Free building work is not merely the expedient provision of buildings.

To turn this key, one must look no longer at the building plans alone, but also at the nature of work and of working relationships. It took me many years to realize this. In this sense the journey was much harder than I had imagined. Indeed at the outset, I had never recognized it as a journey – just as a job.

I also have come to realize that every potential that has emerged from this way of working has been matched by a problem. Needless to say I only came to recognize the problems after they had become fully grown. In this way also, the undertaking was harder than I had ever imagined. I have now become aware that the underlying attitudes which may appear so idealistically irrelevant to people who consider themselves practical and down-to-earth, are in fact of fundamental practical significance. Had I held this view at the outset, how much easier and more effective the whole undertaking would have been!

Once we recognize that architecture can be an art, we come up against the fact that on the one hand buildings are built by people and are homes to people. If, as commonly happens, they are

---

* I use the definition: that which raises matter to the spiritual. A work of architecture, sculpture or painting is still only an assemblage of certain pieces of brick, timber, stone or pigment – yet their intention, given form by this assemblage, can work deep into my soul and touch my spirit so that I can recognize that in some way I will never be quite the same again – my spirit has been extended, as it were. The vibrations of music and physical movements of dance likewise remain material definitions, but can be imbued with the spirit to become art. This definition excludes much that goes under the title of art and includes much that occurs in the context of everyday life.

8

excluded from the design process, the end result will neither be fully satisfactory to its users, nor be imbued with artistic quality by its makers. On the other hand, art, although it may emanate from a universal source and aspire to touch the chords of the universal in every person, is the product of an individual attunement (though sometimes by a closely knit group). There are dangers here. Projects which, while durably founded upon community, allow a situation where everyone wants a say but no one does anything – or worse, anyone can do his or her own thing – can end up in chaos. Yet those that hinge on the dynamic of a single individual can easily turn to a form which exploits the labour of others, while excluding them from a more meaningful contribution.

A shared inspiration and commitment to work together is essential if the elusive middle road is to be found. If it can't, the results can be disastrous – perhaps even mean the demise of the initiative. Sadly, such collapses of ventures, full of good intention, happen all too often.

In every undertaking, money, energy and duration are bound together as three variables in a single equation. In contract building, energy multiplied by duration costs money. The equation can be simply resolved in solely financial terms. Aesthetics or craftsmanship can likewise be reduced to their monetary consequences and evaluated on that basis.

In a gift-work situation, energy and duration do not cost money. Released from the shackles of money, work is allowed to unfold as an artistic activity, allowed to find its rightful role in society as service meeting need, as gift. Where we have become adjusted to the norm of work *sold* for money, we can now stand back and see that work given is a gift to all. It can and should be both service to the recipient and fulfilment to the giver.

To fulfil these aims requires a complete reappraisal of our accepted approach to work and working relationships. This is the potential and the challenge.

But volunteer work also has a price – a price in terms of energy and duration; sweat and patience. Too easily this price can grow to destructive proportions as exhaustion and frustration.

As a job progresses, the critical limiting factor may shift between money, energy and duration. Sometimes, whatever the delays or drudgery, there is no alternative to the cheapest materials, methods and sequence: second-hand bricks, hand-mixed foundations, trickle cash-flow. At others, however, speed of completion may be so important that if it can't be attained otherwise, borrowed money has to be spent on employing contractors or on temporary buildings. The savings made earlier disappear overnight, leaving only their costs: delay and drain on energies. Had changing situations been recognized in time, a little more money spent at the right times could have saved a great deal of frustration or expense. It is not easy, however, always to recognize sufficiently ahead of time how the situation will change and adjust policy accordingly. Here again, it is the foundation in shared inspiration which gives the will to overcome such potentially divisive pressures.

I began with the desire to do things with my own hands. This progressed to a realization that to be fulfilling and meaningful, work must essentially be artistic. I eventually began to understand that to function effectively, the forces at play beneath the surface of perceived reality must be understood. In harmony, these forces can nurture the unfolding flower. Left to themselves they can tear any initiative apart.

From the experience of gift-work and the values and relationships that emerge, I have come to realize how far building work can progress from that which is normal in the work around us. This norm is aptly expressed in the phrase 'Get in, get out, and get paid'.

If work is regarded as gift, as the sowing of the seed of spiritual inspiration into the world of matter, as the redemption of material substance through art, the values of the world around us are stood on their heads. If we take time and observe the headlong rush down the slope of materialism, towards economic, ecological and social disaster, and negation of the human being, we can only conclude that the values of the materialistic world *must* be stood on their heads.

The process of involvement with gift-work building had led me to this realization. It is also one way to set these opposite values in motion.

This has been my journey. I have found it exciting, demanding and sobering enough to feel the need to share it. I have the hope that my painfully earned and severely limited knowledge can become a stepping-stone for others to carry the work forward towards the true, the beautiful and the good.

II

PROJECT HISTORIES

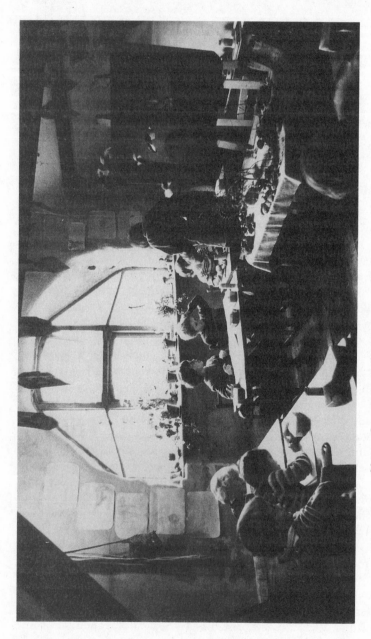

Nant-y-Cwm Steiner school: a classroom in the new upper floor.

2

Nant-y-Cwm Steiner School

IN A certain sense, Nant-y-Cwm Steiner School was brought into being by the building. In 1977 a group of parents and friends in rural west Wales established a Saturday playschool to bring to children some of the aspects of Steiner education – experiences generally not otherwise available to them. It was held in a private house, but the owner became ill. The next house was only available until its owners moved. We decided we should look for a more settled home. Local schools on Saturdays were too expensive, village halls often in use.

One of us came across a dilapidated school building for sale. Like many Welsh rural schools, it was located in the centre of a dispersed community, in other words in the middle of the countryside, about five miles from the nearest village. Depending on the weather, visitors these days experience the school as in idyllic landscape or lost in the middle of nowhere. I first saw it in November drizzle; gaunt, empty, unloved. It would have made an archetypal haunted house for a horror film. Inside was no better. Vampire slogans were on the walls, part of the floor had rotted away. From the next-door farm (a mile away) we borrowed a ladder to make an inexpert structural survey of the roof timbers. Crawling through the roof space I warned my colleagues to step only on the major ceiling joists. Just as I was inspecting some particularly interesting rot, I heard a crash behind me. Of my two colleagues, my torch could find only one. Slowly the head and shoulders of the other emerged from a hole which he had just stepped through, sending a ventilator and nearly himself down to the backbreaking jumble of broken furniture on the floor fifteen feet below. It was an inauspicious start.

The building was clearly quite unsuitable for the Saturday school, yet with enough work, could have been suitable for a proper school, but this was quite out of the question. We would all of us have liked to see a proper Steiner school in Wales, but without money, teachers, any experience in organization – or even any guarantee of pupils – how could we even entertain the possibility? I remember thinking such an idea so ridiculous as to be quite mad – and I said so.

In due course we bought the building. Between us we scraped up the necessary £6,500 and in the spring of 1978 we became owners of a derelict shell, empty for the past twenty years and in urgent need of repair. I still thought the project was mad, but none the less the right thing to do. We were a small group, about ten, although there was a lot of interest from other people. In time we expected to start with twelve children. In fact it turned out to be three, although this number quickly grew, with five others we had not previously met.

We held a meeting in the gaunt cheerless building – some fifty people came and many wrote down their names and ways they could help. We started out with a work day. Each person brought materials and tools. As well as clearing away rubbish and old cars we tried to arrest further deterioration of the building. There were some novel methods such as gluing patches of loose slate together. An invaluable contact, a builder who actually owned a cement mixer, came on the wrong day. After waiting for some time he got so fed up with everyone else's unreliability that he went home and we never saw him again. Except for odd jobs, we could not start any major work until we had building regulations consent. In the meantime enthusiasm seeped away.

From the first, we were committed to consensus decision-making, but we had no formal organizational structure, other than appointing a treasurer. I offered to look after the building and design side. Though there was no formal appointment and no instructions, there were implicit assumptions, basically that everything should be as cheap as possible – if not cheaper – and that the existing unprepossessing building be transformed to be fit for children. We wanted to start as soon as possible so had to use as much as we could of what was there already, however limiting. But first it should be made into an appropriate environment for the education of the children. Along with improvements and structural repairs, there was the practical need for more classroom space, as Steiner education requires activities for the children involving movement, rhythmical exercises, and artistic activities which differ for each age and require separate classrooms, even with only a few children in the school. This we achieved by putting in an upper floor. We also needed water, electricity, toilets and drains. (The existing toilets were holes in the floor of an outbuilding, located over a conduited stream routed right beside the well, the only water supply!) As a second phase, we planned to add a classroom each year. Although I didn't realize it at the time, these three requirements: economy, appropriateness, development of the existing situation, mutually reinforced each other.

Absolute economy meant that we had to do everything ourselves – and that meant that time was freed from monetary cost. We could spend time on hand-work which no contractor could ever afford.

Educationally, the building should provide an environment nourishing for children. In no way should they be forced into a mould by the architecture, but should be free to move, live and imagine in their own child world – to be growing individuals. Any imposition of standardization would negate this. The more individual the atmosphere of each room, and indeed the more individual each situation, each meeting between architectural elements, each element – even each door handle – the better. This meant hand-work, and had the work not been voluntary it would have been financially impossible.

Using the existing rooms meant transformation of that which was already there. The rooms were high, with windows too high to see out of. They were rectilinear, sterile, unfriendly, and above all institutional. To de-institutionalize them meant not only bringing them down to child scale, but bringing life into them. Life depends upon breath – a fluid tide moving back and forth between polarities.

Bringing conflicting elements into harmonious conversation*
brings breath and life to places. This in turn, means that every
situation must be met and resolved individually, without recourse
to stereotype. This sort of thing cannot just be described and then
done by someone else. The person actually doing the work has to
have some feeling, or the result will be very much that of a formula,
sometimes so inappropriate as to look contrived and out of place.
Without work done by hand, dependent on gift,† without a
commitment to individuality and to conversation – both in the
material building and the people working on it – these aims could
never be achieved. With them, we could build a quality into our
work, that money could not buy.

Even when we had got all the necessary permissions, there was
no money to start work. Building work costs money. In an area of
low income, most of the group had significantly lower incomes
than the national average, and those who had been able to support
the purchase had no more money to spare. Confident that we were
embarking on a worthwhile project, we sought donations from
further afield. Rightly – so hindsight shows – our confidence was
not shared. We had at this stage only a dream without substance.
We received no donations. The project was a stalemate.

After about six months it became obvious that we were faced
with the choice of starting work without any money or putting the
building up for sale. There was so much work to do organizing for a

full-time school that we couldn't keep on with the playgroup.

* By conversation I mean the coming together of two or more voices,
people, building elements, or whatever else *to create a single whole greater than
the sum of its parts*. This can only be founded upon listening to each other
and to the theme that is coming into being. One's own personal baggage of
ideas, opinions and preferences must take second place.

† I intend quite specific meanings in connection with giving and taking. To
give: not to hand down or impose but to offer. This is an outward, selfless
gesture, nevertheless requiring a strength of individuality. By contrast, to
take is not to receive or accept but to grab, purloin, exploit – an inward,
egotistical gesture. Egotism and strength of individuality are often
confused but are really quite distinct. One is fuelled by desires – it is a kind
of weakness; the other is the strength to resist such pressures.

Although non-profit-making, it had, by a mistake we have never been able to repeat, made a profit of £36, which it voted towards the school, and with this sum, insufficient to open an account at a builders' merchant, we started work.

Of necessity, we started on work which did not require the purchase of any materials – namely hand excavation. Initially two of us started working two days a week. As time progressed, others came to swell our numbers, small sums were raised by barn dances, craft and produce stalls and the like. We were given a day's JCB time to excavate the cesspool. Then the winter rain started. Each day that we returned to the cesspool, the water table had risen so that we were continually working in deep liquid mud, assaulted by the noise and exhaust of a hired petrol pump beside us in the hole. Day after day we thought that we had built the walls above water level only to find the next day that it had risen.

When the cesspool was nearly full and the recent ice had just thawed, one of us steadied himself on a newly laid concrete block, dislodged it and fell in. The block followed him, fortunately missing his head. As we consoled him afterwards, it was also provident that it was only of water that the cesspool was full!

No sooner had we dug the drainage trenches, than the frost was too severe to lay the plastic drain pipes. Work at this stage was heavy, dirty and unredeemed by skill or attractive results. We were generally bogged down by winter mud, rain and subsequent abnormally hard frost and snow, with nowhere warm to shelter from the weather. It was redeemed by the excellent spirit on site. I became particularly aware that working together can transform the most burdensome drudgery into an enjoyable experience.

As we worked, others came to help us, some on an occasional, some on a regular basis. The sales of crafts, handmade toys, home produce and the like began to raise money sufficient for us to buy our first building materials. And then, quite unexpectedly, we began to receive donations. They ranged from the odd £1 to £1,310 from an anonymous donor. They included donations from USA and Germany, Holland and Switzerland, and from fundraising efforts from people in other parts of Britain, who had no contact with the school. Together with loans of £2,000 that we were able to raise, the money was enough to cover our building expenses. But more than this, the encouragement we received from these donations, small as well as large, many from people whom we had never met, was tremendous.

Similarly, amongst those who came to help us were several who had no vested interest in the school. On some days even, the whole workforce consisted of people without children who might benefit. Their work was clearly gift and not investment. It was our first lesson, that once one steps from intention into deed, if it is the *right* deed, support will come at the right time from the most unexpected quarters.

While the building was struggling along, the school itself was coming into being. We now had teachers. One started the kindergarten class in a nearby house in January 1979. Two others were completing their training. The date when the building had to be habitable for the school to open had to be set. We chose the latest date possible in September.

Under pressure of urgency, the workforce swelled. Six to twelve people each came one day a week, some even working several whole weeks. None the less, with the amount still to be done, it seemed hardly possible that we could complete the work on time. We decided therefore to employ one parent who was a professional builder. He worked for a few days and then found more profitable work elsewhere. A ton of plaster, ordered on the assumption of more rapid progress, became spoiled in storage and had to be thrown away. In desperation, we decided to employ two plasterers for the classroom ceilings that had to be ready at the end of the month. They didn't turn up on the agreed date. Eventually they came, and completed their work three days before opening day. The atmosphere was not very good. To start with, the professionals criticised the amateurs' work and kept themselves apart in both work and meal breaks. The volunteers resented the sum paid out to contractors. It was about 12 per cent of our expenditure so far but appeared to be for only 1 or 2 per cent of the work. By the end, however, the two groups came to be reconciled to each other.

We realized that the combination of professional-paid and amateur-unpaid labour is potentially very uncomfortable. Had we had money, it would have been better spent in the early, uncomplicated stages, by hiring a JCB for all the excavation, and even perhaps block-layers for the straightforward blockwork. It would also have been easier if volunteers had followed on from professionals rather than the other way around. Pressure of time however had become so acute that we had to resort to unthinkable expenditure. We spent in the wrong order and it cost us more, and more time and effort, but to spend the right sums *at the right times* – especially when the money is not there – is by no means easy! And in any case, the money was not there.

Even in the last two days before we were pledged to open, it seemed impossible that we could complete on time. In addition to our own unfinished work, walls, windows and floors were covered with plaster droppings. Scaffold, furniture, tools, oddments and rubbish were all cemented together with it. None the less on Michaelmas day 1979, after an expenditure of £4,239 and ten months' work we cleared the last building refuse out of the classrooms and were able officially to open the school, albeit in a rather unfinished state.

The larger classroom, doubling as hall, was divided by a temporary and acoustically hopeless partition. The walls were bare plastered and much else was unfinished. Beyond the two classrooms was a building site, through which a way led to unceilinged toilets with rough concrete floors, their roofs under-felted but not yet slated. Outside was scaffolding, rubble and builders' mess. It was not satisfactory, but it was just *passable*. The school was therefore *possible*.

What had been a dream, a vision, an inspiration, a task, had now become a reality. The building could now find its proper place, no longer as the absorber of all energies, but as the material home for the spiritual being of the school. When we looked back to that apparently lunatic decision to purchase the building a year and a half previously, it became clear to us that our actions in the material world had enabled something spiritual to come into being, something that was waiting for us to start to act.

I am reminded of Goethe's words 'Whatever you can do, or dream you can, begin it. Boldness has genius, power, and magic in it.'

Of course, other, new, problems lay ahead.

THE SCHOOL was now in operation. Whereas previously its existence had depended on a space, now it depended on money. Money-earning activities such as craftmaking and cooking began to absorb much of the energy that had gone into building. Few of us had anticipated how much time and money bringing children to school would take, there being no public transport anywhere near. Free time hardly seems to exist for those with self-reliant, subsistence life-styles and any gift of time at all represented great sacrifice. Moreover, building was now no longer seen as a priority! Building was for the future and few turned their eyes beyond the struggles of the moment. More urgently, however, the temporarily divided classroom was difficult to teach in and the building was not insulated against the approaching winter. Some doorways were only curtained.

Volunteer numbers had fallen until I was often on my own and it became clear that the essential and urgent work exceeded our capacity within the time available. We decided to pay one of the more regular volunteers one whole pound an hour for three days a week, for a month – barely subsistence wages!

One person on his own is at his least effective and although he appeared not to get demoralized, I often felt overwhelmed by the immensity of the work ahead of me. For everybody else however, especially the teachers who had to live with the unfinished school, work made a big bound forward. Although still slow, it was now three times as fast! Attempts to arouse a sense of urgency about building were met with suggestions of temporary solutions such as renting Portakabins or rooms elsewhere, buying ex-army sheds or borrowing caravans. Attempts to cajole people into volunteering their time, succeeded only in creating 'them and us' and 'they should' resentments.

Our next task was big. All we had of an upper floor were the joists with a ceiling beneath them. When in due course we were laying a floor on these joists, inevitably someone trod on the unsupported plasterboard. The children below, looking up at the sound, saw a foot protruding through the ceiling.

To use any of the upstairs required raising what had previously been the kitchen annexe (built in the fifties – just before the school closed) and breaking into a large area of roof over occupied rooms.

This we had to do to get space for the necessary two staircases for escape in the event of fire. This roofing work had to be carried out as quickly as possible as tarpaulins are not trustworthy in Welsh winds. Only after the roof would we be able to turn our attention to flooring, putting in windows, partitions and ceilings and all the other work involved in making classrooms in what was about to become 'upstairs'.

With spring we had to start on the roof or miss a year – and what would happen when there wasn't another classroom? Fortunately more people came to help, and as the job and its importance became more and more visible, so numbers increased until, when we reached the highest point, we could hold a topping-out ceremony with six of us, together with three teachers and nineteen children.

We learnt that the gift of time seems to come in wave patterns. Although perhaps a little later than one would like, clearly *apparent* need is met by help. When a job *appears* to be more or less finished, the sense of urgency wanes and so do numbers of volunteer workers. Starting something new brings new enthusiasm, especially when it reaches the stage where real changes are visible. Starting a new phase of continuing work does not elicit the same response.

The troughs are difficult. Because it is so very hard to restart work that is allowed to peter out, it must be kept going. At times, the atmosphere on site becomes weighted by feelings of being taken for granted, exploited and undervalued. On the whole, however, the morale was good.

The composition of the workforce varied. Sometimes it was all men, sometimes all, except myself, women. Sometimes Welsh, sometimes English, sometimes European. Sometimes work went smoothly, sometimes not. With changing faces came problems of continuity.

In the electrical trade, for instance, it is common for each electrician to have an individual notation for marking wires. Furthermore, there are several systems by which wiring can be laid out. Electricians don't like to take over other electricians' systems and notations. Over a period of six years we had six electricians. Each promised to leave comprehensive notes and wiring diagrams, but each left the site more suddenly than either they, or we, expected. Consequently, each electrician spent much time checking, and often ripping out his predecessor's wiring. The building was never at a stage where any zone of electrical installation could actually be completed in one go. In one case, an electrican, knowing he was about to leave Wales, laid cables for us in an area where we had not yet finished demolition. In due course they were damaged without anyone noticing it. During the holidays a visitor looking around happened to notice smoke!

Continuity can be bought, but this also has problems. When someone started offering wages for help in erecting an urgently needed temporary building, it came back to me as 'How come so-and-so, who has nothing to do with the school, gets £20 a day while we work for nothing?' If urgency or need of specialist skill requires having to pay for work, I always look first among people who have given time and effort. If, thereafter, they are lucky enough to be paid for their work their attitude will still be in sympathy with the gift-work ideal. In such cases wages mean a contract of reliable attendance. The end of free work, but not the end of gift-work. Rates are always well below contractors' rates – and that is important, for I want people to choose consciously to give when they work, not to try to make a profit from it. In return, their living needs are met.

Sometimes someone has paid for an outsider – usually sympathetic to our aims and situation – to undertake a particular task. It is very heartening to arrive one day and find a new job done. But as these outsiders preferred to fit in odd jobs when convenient rather than commit themselves to specific dates, supervision was impossible. There is still minimally lapped slating that I would like to see re-laid, although miraculously it doesn't seem to leak.

On a number of occasions, we were promised work by people who had travelled great distances to help us. They wanted to start straight away and so it was necessary to arrange a meeting on site, perhaps not on a day when I would be working there, to go over the job carefully. All too frequently they never reappeared. This makes a difficult situation. People have time to give. Their offer is, I believe, *always* genuine, but their reliability is unknown. There is no unlimited fund of time (or materials or tools) at the receiving end.

24

Everyone, supervisors included, is giving his or her scant time. One arrangement suits the regular volunteers – in our case, it seems easiest to squeeze one day a week out of the demands of daily life. Another pattern – blocks of work – better suits outsiders. We have also tried work days and work weeks, limited to certain tasks, such as finishing off an area. Generally work days have been effective and good fun. Work weeks on the other hand have only succeeded in completing work that could have been done in one day, but by dispersing the workforce have created demoralizingly lonely and often ineffective working situations. Working weekends have been a good compromise.

The pattern of work that emerged as most suitable at Nant-y-Cwm is, of course, particular to the situation. Elsewhere, other patterns – perhaps block periods, work days, evening or weekend work, or work organized around defined projects or individuals, may prove more appropriate.

Co-ordination has always presented problems, particularly when one person takes over where another has left off. It was all the more acute when individuals whose desire to do their own thing was so strong that they avoided regular working days. Sometimes one group has brought work to a certain stage with great sensitivity. It was then taken on and the whole direction changed by someone else. On one occasion somebody ripped out half-completed built-in furniture so as to replace it in another style. Another chose to introduce a Victorian gothic aesthetic to his window. In both cases the individuals shunned discussion or departed from agreements. This can create a delicate situation. On no account do I wish to spurn an offer of help, yet it is often very difficult to distinguish, early enough, an offer of help towards a common aim, from one that is going to turn out to be an individualistic but perhaps disharmonious contribution.

Various similar problems have arisen through lack of contact even where the intentions have been good. These problems and disappointments must, of course, be seen in the general context of success and good spirit. Indeed, most people find work a nourishing experience. They come because the work needs to be done, but they keep coming because they enjoy it. When a television team came to film the school, we phoned around to get

25

as many people as possible to come to build. We had several mothers who had not been before and who enjoyed themselves so much that they wanted to come again.

One day someone came to us, I think out of curiosity and having nothing to do. His life was without direction, he had no skills or attunement to building and I was continually worried that he would have a serious accident! When he started he was barely able to use a hammer, yet as time progressed he became confident, capable, responsible and enthusiastic. The experience opened new inner and outer horizons for him and when he left us eighteen months later he went to look after handicapped children and is now training in therapeutic sculpture. His experience made me realize that work has a vital inner purpose – it is not just for achieving production!

IN OVER nine years of building, the school has grown. New individuals have replaced the old, bringing with them new perceptions, assumptions, viewpoints and ideas. At times this gave rise to problems.

The original group were essentially pioneers – to them economy, improvisation and an incompleted environment were unavoidable – an inevitable consequence of things going forward, almost to be enshrined as a virtue. Some of those who came later had the distance from events to be able to observe that while such a situation may have been and perhaps remained to some extent the outer necessity of circumstance, it did not necessarily serve, and in certain ways hindered, the real purpose of the enterprise.

For the building group, work had become the task of building an environment fit for a school, artistic and health-giving. The teachers on the other hand considered that the children were not best served by an education in the midst of a building site.

Builders also had a sequence of work in mind to protect the building from winter and to keep jobs open for occasional influxes of unskilled helpers. The teachers wanted space. Space became an area of conflict – the *order* in which it was finished and *who* used unfinished space. One party wanted space to work in, the other to store school furniture which would soon be needed.

Eventually we were able to agree a new set of building priorities, but not before a bitter division into 'them and us' had arisen. It was one that took several years to completely heal.

None of us had been prepared for what had started to happen. On one side the design, method and programme of building had developed its own momentum, evolving according to its own requirements. So had the educational stream. Both were so busy that they did not concern themselves with the other until they needed to, in other words until there were problems. Suddenly, like a family crisis, what had been quite unnoticed and apparently normal one day, became overnight a threatening rift. Specific focuses of work had led to separate streams with inadequate communication. Each stream had formed its own priorities which it then advocated as the only possible way. Both parties lacked flexibility, but what we all needed, above all, was an attitude of continuing, listening, conversation. How, in retrospct, could it best have been achieved?

It may sound trivial, but it is a tremendous help if everyone has tea, lunch and other social occasions together. In any institution, this is the best time to cross departmental boundaries. It helps to have clear priorities, agreed by all those involved, and not just be at the whim of daily events. Writing these down gives them more substance than does verbal agreement alone. Many people tend to remember points in a debate more vividly than the decisions that ensue.

Priorities and programmes, like design, need to be flexible. However sound the original arrangements, organizations evolve, and new people come into them. The situation is constantly changing. For this continuing evolution to be a source of benefit rather than of conflict, as many options as possible need to be kept open as long as possible.

Sooner or later, however, decisions are made which finalize intentions as to use and appearance. Building – and heating – these days are so expensive that I am of the opinion that every cubic inch must have a purpose. This implies very tight design and, if the client will support it, tight space-management, so that most space is in use most of the time and is matched to a heating regime, for instance waste heat from a pottery kiln. Space is too expensive to be under-used. Space-management, like energy-management, must however be seen in the context of more holistic environmental considerations. Some spaces need to be dedicated to a particular mood, indeed to a particular spirit, and when not in use for this specific purpose are better lying idle.

We used, for instance, to hold our evening trustee meetings at Nant-y-Cwm in the kindergarten. It took an experienced teacher to point out to us that the spiritual echoes of these sometimes contentious debates eroded the warm, secure magical enchantment that the kindergarten should provide, and affected the children adversely. We realized that the conflict of spirit was as damaging for the mood of the room as if a church were used as a night-club on weekday evenings. Both inside and outside buildings, many places need to nurture their own particular spirit of place. Such a pattern of restricted space use is conscious, but more commonly, space is under-used simply because better use is inconvenient.

There are so many constraints in conversion work that there is rarely any room to spare for loose fit. This is especially so where it is an essential functional requirement that the environment is an artistic whole. For a Steiner school this is definitely the case. Both for functional and artistic requirements, the environment is either appropriate to its use or it is not. Rooms for a lower or an upper school may have the same size requirements, but adolescents are quite different from children in the flower of childhood. Their education is accordingly different and so also should the architecture be – firmer shapes with stronger axial orientation for the older, softer and more fluid ones for the younger. Somewhere between flexibility and precision lies the right balance. They are not quite incompatible, as flexibility relates to use and precision to the qualities of the space that is used. But I cannot say that I have ever found the exact balance.

BUILDING WITH volunteers is meant to be cheap. Every situation will be different, but a cost picture of Nant-y-Cwm can give an idea of relative expenditure.

NANT-Y-CWM STEINER SCHOOL BUILDING COSTS

Year   *Expenditure*
        £

Year	£	
1979	4,335	ground floor usable
1980	3,924	
1981	1,111	one first-floor classroom usable
1982	2,535	two first-floor classrooms usable
1983	2,534	
1984	1,897	
1985	2,611	upper floor complete (less a myriad of minute
	18,947	finishing-up jobs forestalled by occupation)

    cost per sq.metre: approx. £81.60 (including VAT)

method 1:

total area	291 sq.m.	
construction cost	£18,947	
cost of existing building	£ 6,500	
	£25,447	
cost per square metre:	$\dfrac{£25,447}{291}$	£87.47

method 2:

allow ⅓ for existing shell of	122	sq.m.
to be completed ⅔ × 122 =	81.2	sq.m.
and new floor area	169.1	sq.m.
	250.3	sq.m.
cost per square metre:	$\dfrac{£18,947}{250.3}$	£75.70

By comparison, I am told that the average price for independent schools at that time was £550 per square metre + VAT = £632.50 per square metre.

By using gift-work and second-hand materials, Nant-y-Cwm was built at between 12 per cent and 14 per cent of estimated contract costs.

HAD WE been more efficient, particularly in timetabling and looking after building materials, it would have cost less. The financial picture is attractive, but we must look beyond it to make any meaningful evaluation.

There are numerous minor defects in the construction, though none is in my opinion serious, nor are there significantly more than I find on contract sites. There was intense frustration over the time taken, but none the less no class or activity has ever had to be cancelled due to lack of space. There was the problem of reconciling the mess of a building site with the ordered form of an educational environment, eventually and belatedly resolved.

Were it not for the low level of expenditure, spread over several years and of such a scale that every contribution of money or work could be seen to make a difference, there would be no school. What was previously an idea – a hope, without a ground – is now a being with a life, an evolution and a magnetism of its own. The experience of working together upon which the school was built has developed into a strong community.

Opportunities for hand-work, evolutionary design, and conversation between what was and what will be appropriate for the future could be taken up in a way impossible with contract labour. This way of working has great potential – in the aesthetic, economic, social and personal development spheres – to enable, to give birth, to heal.

At Nant-y-Cwm, we have unlocked only some of that potential, but in our view it is worth much, much more than the sum of all the problems we have had to overcome.

At a building conference: people from many nations shaped this ceiling. Working together with the hands can overcome problems of language and build deep friendships.

3

Volunteer building conferences

THE IDEA came from a Norwegian architectural student I met at a conference in Sweden. 'Why not have a conference in which we spend our time working together?' 'Why don't you arrange one in Wales?' asked a Swedish teacher. We did.

In the course of arranging the conference, the members of the organizing group differed widely in their perceptions and expectations. Amongst the suggestions were a host venue for a group from Vienna who could lazure-paint the walls for us with expensive natural paints*; a work camp where we would offer space to pitch a tent in exchange for labour; an assembly of up to 200 people participating in some fifteen craft and discussion workshops. The main body of opinion lay somewhere in between these extremes.

However, as we worked towards the point when we had seriously to launch the conference or forget the idea, the uneconomic, egotistical and impractical proposals fell away and we realized that we could only offer that in which we were experienced, which we had the resources to mount, and which

---

* In lazure-painting a number of extremely thin veils of translucent paint are applied over a textural white background. The light is reflected differently by each layer of paint so that the colour appears to live, hovering about and through the surface of the wall, ceiling, etc., rather than lying rigidly 'dead' on the surface.

contained sufficient benefit for participants to justify their attendance. The themes that emerged at that time in a rather loose form were those which subsequently we have been able to develop more clearly and consciously.

Working together would be a central theme, leavened and set in context by artistic and cultural activities. As such, we intended that both the conference as a whole and the practical work within it should involve the whole human being – head and heart as well as hands. We were anxious to balance what we received by others' hands with that which we gave. Building work, therefore, never took up more than half the time, nor was it just *any* work. We chose jobs that could be worked on by a group, and that would, in a brief week, show marked artistic results. This meant preparing jobs to a 'threshold' stage. At the beginning of the week the chosen work area felt like just a building site. By the end, it had become an ensouled space – a beautifully shaped or coloured room, a garden, or a woodland grove; something with which one has developed a relationship of feeling; something which the experience of working on was personally rewarding.

We have to plan our work several months ahead to have suitable jobs prepared to the right stage. They must be suitable for group work, fulfilling to work on, regardless of personal skill, and accomplishable within the time available. We look for jobs which are uncomplicated for teams of four or five to start, even if subsequent tasks will require single individuals. This is easier to manage and gets us all off to a good start.

Involving thinking and feeling with doing raises labour to creativity. This threefoldness, repeated in the conference timetable, helps to nurture the whole human being – not just the head or hands as in many conferences or work camps. The timetable balance between giving and taking creates a breathing relationship between guest and host. That which we give, we also enjoy; that which we receive – the gift of work – is experienced by those who have attended the conferences as our gift to them. The threshold nature of the jobs makes visible more readily in the short time available, that we work together to transform matter from its raw and unattractive state to something nearer the artistic. If we think of art as that which raises matter to the spiritual, we realize that we are struggling to find a way to do just that.

Although less clear then than today, these intentions were all present from the start. The first conference in 1982 was attended by twenty-two people from six nations, including the six of us. We concentrated our indoor work on plasterboarding the complicated three-dimensional surfaces that would enclose a classroom out of an empty roof space. Outdoors we made a stream. The classroom was transformed from open studwork and exposed rafters to a room, revealing its full complexity and richness of shape. The stream had formerly been conduited on an artificial route underground. We brought it to the surface to breathe in the air. To do this, we had to shape the land in an appropriate way, so that the stream was not an alien, but could belong in its place. We had to create a conversation between the stream and the land. And then the water wanted to sing! We manoeuvred rocks in its bed till the sound seemed at its happiest.

When first we let the water flow, the stream did not sound happy. It sounded squalid and dirty – as indeed it was. In an hour or so the sound had changed; the stream sang. It is hard to describe it, but the water *sounded* clean. When we looked we saw it was.

Subsequently the stream became a magnet for the school-children's civil engineering. We realized that we had failed to take into account the fact that a natural stream has its own formative powers which can accept a reasonable degree of manipulation and play. An artificial stream is much more fragile. Instead of working with the stream itself, it would have been better had we worked more with the landscape forms which bring streams into being, regardless of how children dam or re-route them.

We offered eurythmy* and a choice between painting, projective geometry, and social exercises. Lectures brought into relationship the ancient Celtic past, the present, and our spiritual tasks in the future. Outings reinforced the lectures and provided good relaxation and enjoyment, finishing at dusk on the beach. In

---

* An art of movement which makes visible the normally hidden, formative qualities of speech and musical tone and interval. Inaugurated by Rudolf Steiner and Marie von Sievers.

planning the conference I had regarded food as a necessity, and little more, but the food that appeared on our miniature infants' tables, dressed with flowers and set out under the trees, was a special point in the memories we took away from the week and a lesson to me, even if well known to other conference organizers.

We modelled subsequent conferences on this successful format, and a pattern began to emerge; the indoor work was of a spatial quality and brought something into being; the outdoor was more in the nature of creating a relationship – a conversation for instance between sunlight and shade where, by selective felling and trimming, we tried to achieve a delicate balance between accessibility for play and the magic of woodland. In other years we worked with the conversation between building and ground, with brick-paving, wall-painting and planting.

BY 1985 our outdoor work had moved into the landscape, bringing landform, garden and building together with dry-stone terrace walls, subsequently further developed by a 'gardening with young children' conference. Indoors we transformed a harshly cuboid room, built in the 1950s but now after our extensions above with exposed timber overhead and gaping holes in the walls. We fitted windows, shaped the room and plastered it. Two other classrooms with contrasting sunlight conditions and for markedly different ages of children we lazure-painted. We had now, in miniature, been through the whole of the final process that brings architecture to life before it is set *in* life by occupation and use. We had worked with shaping the boundaries of space, with textural surface and with light and colour. These things tend to be regarded as the inessential frills to architecture, but in fact make a fundamental difference to how we experience it – as approachable or unwelcoming.

This time, for our third conference, there were eighteen participants and ten of ourselves, representing nine nations. The weather was generally poor, but the tempo of the week in both work and other activities was intense. We completed much more than I had ever anticipated. For myself, and, I think, for others, it was a watershed experience. The world (or perhaps it was ourselves) seemed different after that week.

As we gained in experience we were able to refine our organization, timetabling activities very tightly in the early days and, as people have more and more to talk about as the week goes on, leaving progressively more and more space in the timetable towards the end. We have never quite resolved the problem of how to fit everything into one week. We want to offer both a complete plan of lectures, trips, artistic activities, discussions and recreation, and a sufficient length of time for the work to achieve fulfilling results. To add to our difficulties, we feel committed to the policy that the work should not exceed half the time. Even though we realize that it can be the more rewarding part of the week, we feel it is not right to take more than we give.

The appeal of the conferences seems to vary widely from year to year. In 1987, flushed with previous successes and feeling that we had too much to squeeze into one week, we developed two distinct themes for two weeks. The first, concentrating more on building as an artistic experience and on architectural issues, the second on listening to the environment, most especially to that realm where different elements such as rock and water, tree and ground, light and shade meet. Such meeting points are where the invisible life in nature is strongest, and is nurtured. Observation here can distinguish whether a landscape is healthy or sick, and sensitivity to what is happening can help bring landscapes back to health. These we thought were important themes to the needs of today. None the less, we had only four bookings from outside our local circle of supporters.

For all the fluctuation in support, the conferences are of benefit to us, the organizers. They give an opportunity to define form and emphasize themes that are barely possible, or visible, in the more unpredictable and pressured atmosphere of the volunteer work-site. Hence the conferences have nourished us, not only by their material achievements, but also by clarifying our understanding of our inner task and our sense of purpose.

They provide a primitive form which, in my opinion, could be adapted to many other circumstances of volunteer or even more conventional projects. Other places have something unique they

NANT-Y-CWM STEINER SCHOOL	**1985 INTERNATIONAL CONFERENCE**											
											Monday 22 July	7.30pm Introductions to jobs, tools, etc. Welsh music, tea, circle-dance
	9.15am VERSE	9.30am EURYTHMY	10am WORK	11am COFFEE BREAK	11.20 WORK	1pm LUNCH AND WASHING-UP	2.15pm WORK	4pm TEA	4.15pm CONVERSATION ON ART APPRECIATION WITH ALAN THEWLESS	5pm ARTISTIC OPTIONS a) dyeing, spinning, knitting b) painting, c) sculpture	6.45pm SUPPER AND WASHING-UP	8pm TALK by Barbara Saunders-Davies OPEN TEMPLES OF THE PAST
Tues 23	VERSE	EURYTHMY	WORK	COFFEE BREAK	WORK	LUNCH AND WASHING-UP	WORK	TEA	CONVERSATION ON ART APPRECIATION WITH ALAN THEWLESS	ARTISTIC OPTIONS a) dyeing, spinning, knitting b) painting, c) sculpture	SUPPER AND WASHING-UP	TALK by Barbara Saunders-Davies OPEN TEMPLES OF THE PAST
Wed 23	VERSE	EURYTHMY	WORK	COFFEE BREAK	WORK	LUNCH AND WASHING-UP	WORK	TEA	FORUM	ARTISTIC OPTIONS a) dyeing, spinning, knitting b) painting, c) sculpture	SUPPER AND WASHING-UP	WORKSHOP, with Gilbert Lamont MONEY - A THREAT TO OUR SOCIAL ORDER?
Thurs 23	VERSE	EURYTHMY	WORK	COFFEE BREAK	WORK	LUNCH AND WASHING-UP	WORK	TEA	FORUM	ARTISTIC OPTIONS a) dyeing, spinning, knitting b) painting, c) sculpture	SUPPER AND WASHING-UP	TALK by Catherine Kastelitz MONEY - AGRICULTURE OF THE FUTURE
Fri 26	VERSE	EURYTHMY	WORK	COFFEE BREAK	WORK	LUNCH AND WASHING-UP	2.00 pm TOURS: a) Plas Dwbl Bio-dynamic farm b) Architecture, with F.S. Day					TALK by Christopher Day: ARCHITECTURAL POLARITIES
Sat 27	VERSE	FREE TIME	WORK	10.30am	WORK	12.30pm CELTIC TOUR – picnic lunch –					Picnic Supper/ WEATHER PERMITTING	
Sun 28	VERSE	EURYTHMY	WORK	COFFEE BREAK	WORK	LUNCH AND WASHING-UP	WORK	TEA	FORUM	picnic lunch – 10.30am CELTIC TOUR	Picnic Supper/ WEATHER PERMITTING	
Mon 29	VERSE	EURYTHMY	WORK	COFFEE BREAK	WORK	LUNCH AND WASHING-UP	WORK	TEA	FORUM	FREE TIME	SUPPER AND WASHING-UP	EURYTHMY DEMONSTRATION GIVEN BY CAROLE NICKEL REFRESHMENTS CIRCLE EURYTHMY

can offer to visitors from elsewhere. Other projects have needs for help and tasks which can be transformed from the necessary for the recipient to the nourishing for the work-giver. All we have done is start, before we quite knew what we were doing.

Nant-y-Cwm conference prospectus 1987

In modern times the increasingly one-sided approach to work furthers materializing tendencies in the world about us and degrades the human being. Seen in this true light, however, work is the meeting ground of spirit and matter; its mission is that of raising matter. It is essential to the wholeness of work that it is imbued with an artistic quality. Without a fusion of feeling and understanding with the practical effort, work can neither be whole and fulfilling, nor functionally and spiritually effective.

Central to working this way is the experience of raising building materials from the purely material to the aesthetic. To experience this in a short time requires work to be organized around two principles: transformation and teamwork. During each course, the building should grow across a threshold, so that, for example, what at the beginning was a skeleton of rafters becomes a space with recognizable aesthetic qualities. To achieve such results and to give creative and appropriate tasks to every individual, regardless of level of skill or strength, we will endeavour to work in teams.

Working together, we can bring into being a social spirit to nurture this experience of whole work, with its balance of gift and fulfilment. The practical work on Nant-y-Cwm Steiner kindergarten will be balanced by artistic craft activities, lectures and seminars, music and recreation. Each week will be equally divided between the gift of work of the participants and the artistic and cultural gifts of the hosts.

These two weeks have distinctly separate themes and may be taken separately or together.

5-11 July: Building as an integrative experience
(This week is particularly suitable for those interested in architecture and building.)

In a student design there is a tendency for spatial awareness to be bound by the limits of paper and model. Building construction is often more a necessary exercise than a reality. Rarely is it an aesthetic experience, something to kindle enthusiasm.

More fundamentally, there is a tendency to compartmentalize the various stages of design – the rational, verbal report, the aesthetic exterior and the fragmentary areas of construction. Yet buildings are integrated unities and to achieve their purpose need designers who themselves can bring an integrated wholeness to their work.

Construction is widely regarded as the stage at which aesthetics stops. In fact, as numerous self-builders demonstrate, it has the potential to transform a mediocre design into a work of art.

For building construction itself to be a nourishing, rather than merely a money-earning experience, it should involve the aesthetic sensibilities and rational understanding as well as manual dexterity and effort. This approach is both enriching for the individuals concerned, and gives a special quality to the building. Building work is no longer drudgery, but creative – an artistic experience.

This week will include visits to architectural projects; artistic exercises and seminars on such themes as:

> Architecture for environmental harmony
> Architecture for physiological health
> Architecture for psychological wellbeing
> The curative effect of environment
> Work with the hands in the age of the machine
> Participants' topics

12-18 July: Listening to environment
(This week is not limited to vocational involvement but intended to be equally suitable to people of all ages and walks of life.)

Right listening is the essential foundation of right action. Failure to listen underlies all forms of non-communication, in architecture as in human conversation. We need to develop the ability to listen, not only to human speech, but also in every sphere of life: amongst them to the landscape with its physical and non-material qualities. We will attempt to develop a listening approach to group work and in our practical work on this sensitive site.

The theme of lectures, seminars, outings and artistic activities will be listening to environment. We will take advantage of the strong elemental life and Celtic legacy of our surroundings.

IT IS through these conferences that the themes which constitute this book, have been refined. And it is the recognition of these themes which has such profound implications for our approach to work in general. Largely it requires a complete reversal of current attitudes. Impractical as all this sounds, we know from experience that it is in fact practical. Nant-y-Cwm conferences have shown that it is so.

40

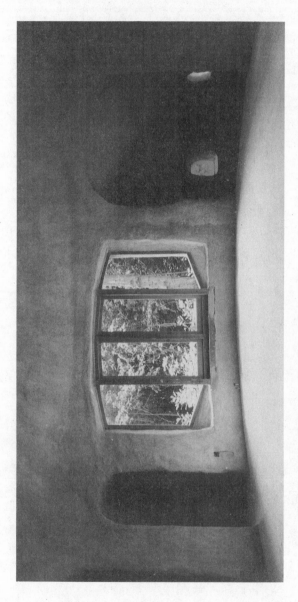

Texture and its interplay with light can make all the difference to how we respond to a room. Life-filled textures tend to make a place more welcoming; lifeless mechanical ones tend to repel. To imprint life into texture means to work with the hand. To find the right balance between the chaotic and the over-controlled, between the dead straight and the over-indulgently fluid means cultivating aesthetic sensitivities in the workplace. No amount of brilliant design in the remote office, on the abstraction of paper, can compensate for the fact that is the people who actually *make* the buildings who can make – or break – them.

(Nant-y-Cwm Steiner kindergarten)

------ 4 ------

A Steiner kindergarten

THE IDEAS in this book have at least coalesced, but mostly been staring me in the face, as a result of many years of practical work. The work formed the ideas. This is the first project in which the ideas have been sufficiently formed to give conscious direction to the work. Some things that elsewhere I found easier to work with have not gone smoothly, others, that I previously had difficulty with, have been easier. At the time of writing, the project is still continuing.

In 1985 I was asked to design a kindergarten for Nant-y-Cwm Steiner School. There was only one site possible, and it was (potentially) ideal – though difficult technically. Building work was to start in April 1986 for a completion target of September 1988. But there was no money! Not until September 1986, when a parent lent £1,000 to enable us to start, could we do so. As concrete for foundations, and bricks, blocks, sand and cement to build upon them would cost more than that, I had to be most cautious with any expenditure.

In the early days about seven people came regularly to build and we had to spend our time on jobs which £100 could have bought – making a site hut out of scrap timber, demolishing the ruined remains of a building, and so on.

Two of us had the difficult task of setting out a complex building on ground complicated by tree roots, ruined walls, and steep slopes. Had we been able to afford a JCB to clear the site, this would have been much easier. In the meantime, with little actual progress, most of the other helpers faded away.

We were given a day's JCB time and promised a tractor with trailer. The latter – although frantic telephone calls established he was 'on the way' – never turned up, so that by the time we could arrange a substitute, the JCB had worked itself to a standstill, covering much of itself, most of the site – and many of the setting-out pegs – with a huge pile of spoil. That we were able to set these out again relatively quickly, showed me how much time we had wasted on an uncleared site.

We were fortunate to have beautiful weather that autumn. There was some rain the day the concrete lorry came to fill the foundations, but we had a big group of people and labour to spare to bale out the puddles in the trenches. I was nervous, though, of the combination of people unfamiliar with building sites and the casual manner with which the driver swung the conveyor boom across the site, at neck level.

Buying concrete was expensive, but I could not face the long drudgery of mixing it – even had we had a mechanical mixer, which we didn't. It was worth it. Before, we just had mud and holes. Afterwards we could feel we had started on the building.

Starting building upwards signalled the end of the good weather. From then on it would be winter rain, gloom and mud. A little more money trickled in and I was able to buy supplies and work two regular days with a regular helper – both of us, for about six months, paid a subsistence wage. We could not alter our hours according to the weather as I never knew when others would come to help us. I therefore had to be regularly there. Some days were wet, but with techniques like covering the mortar in the wheelbarrow and only laying out enough for one block which the other person immediately put on, we were able to work right through the winter, losing only two days because of snow.

From the very first bricks the building has required us to be aesthetically involved. The curves of the walls looked just right on paper – but on the ground . . . First of all, it was difficult to locate them, then – since they were at the bottom of trenches – it was difficult to see them in relation to each other. On the other hand, they were no longer *drawn* curves but now the *boundaries to space*.

For all the care with which they first were laid, I nevertheless had to relay quite a lot of bricks to get the curves just right. There were occasions when somebody, building the wall higher, inattentively straightened some of the subtler curves, but fortunately we were able to bring the shape back at the level where the cavity was filled with concrete.

The curves, combined with the changes in floor levels, ground level slopes, underfloor air vents and service conduits, and so on, demanded a lot of thought. Many of these complications, especially those involving the damp-proof course were easier to *do* than to *explain* and so for a long period my own work was largely working ahead on a section of wall at a time, to free simpler work for other people.

Most buildings look very ugly during the course of construction: damage to surroundings, mess and harsh straight lines. Here, even where the work was fragmentary, the curves were a pleasure to work on; even on my own, and even in the rain, which is more than I can say for other buildings I have been involved with. As far as possible we tried to bring the whole building together, so that the curves could be seen as a rising contour plan. Of course, I then would go and ruin the effect by laying odd groups of blocks here and there to mark out window openings.

Despite the offers of conifer-hating chain-saw enthusiasts, I was anxious to avoid any destruction of the atmosphere of the site. It already had its own mood, its own spirit. *We* after all, are the trespassers. My policy therefore was to fell the absolute minimum number of trees for space to work, or any trees in areas we definitely knew would be open, and any that needed to be thinned anyway. The rest we watched.

We watched to see which trees cast shade where and when. The kindergarten would only be in use in the mornings. The children would be outside at certain known times. Ash trees would only be in leaf some two months of the school year. Conifers, thinned below, could cast their crown shade on the roof without darkening either classrooms or playspace outside.

Where and when did we want shafts of sunlight, selected tunnels of view, to the river perhaps; where did we want the view blocked – of the neighbours' bungalow for instance; where did we want to create mystery and ambiguity? Not everybody believed me when I said that were the site cleared it would shrink to a quarter of its apparent size.

Every so often we cut a few trees down. Sometimes I marked them with limewash, but by the time the promised treefeller came with his chain-saw the marks had washed off and the weather denied us any sunlight effects, making them hard to find again. As yet, however, no one has cut down the wrong ones.

The result of this concern for the surroundings meant of course that we had very little space to store materials – it was the tightest site I have ever worked on. Indeed some materials had to be stored actually within the walls. This meant that organization needed a lot of forethought. Goods had to be ready on site when needed, but not to be there much before then or they would get in the way. We were particular where things were placed: Blocks, for instance, where they could be passed over the wall and chain-handled across the site, or barrowed around paths we cleared. When blocks and bricks had to be moved out of the way, they were distributed in stacks where they would be needed. Mortar could then be moved in barrows along the wall and everything the bricklayer needed was always within reach.

To ease my back, I tried never to lift anything. Stacks of blocks therefore, I did not use up right down to the ground, but kept building them up so that I was only taking blocks off or putting them on between thigh and chest height. Similarly I would try to find standing positions which allowed work to be at the same height. We did not therefore backfill the foundation trenches until the walls were high enough to be worked on from the new levels without bending. In the same way you can build up stacks of bricks six to eight feet away by swinging the body, avoiding endlessly bending and straightening the back.

This was the first site I have run where we had a really efficient system, never double-handling, with goods always flowing downhill towards their final destination. It saved days of work. The demanding spatial pressures of the tight site turned out to be a benefit, as we were forced to be conscious about organizing just what went where, when and how. We just didn't have the room to do it otherwise.

Wherever possible we tried to maximise our efficiency. To lay a concrete floor, for instance, one of us filled the gauging buckets, one loaded the mixer and one spread the floor. The in-between jobs like barrowing, fetching cement or water – or making tea – were taken on by whoever got ahead of the others. This sort of system we use wherever we can. Given that time is such a precious commodity, we just cannot afford to be a 'standing around site'. I have wasted so much time in the past waiting for others or having them wait for me, but here there was hardly any of that. If I had to wait for mortar, I could always replenish my stores or check the site. And we kept the site tidy. This made everything much more pleasant and much easier. Hardly any time was wasted tripping over things, looking for things, cursing mess.

On the other hand there was not as much teamwork as I would have liked. It soon became apparent that my assistant preferred to work on his own. When, therefore, unscheduled unskilled helpers turned up, I had to find ways for them to work with me – but I desperately needed to keep ahead with essential preparatory jobs. It was harder when they arrived halfway though the day when I was in the middle of something particularly complicated. I could always find jobs for people to do, but not always jobs as interesting as I would have liked to have given them.

On one occasion six of the 'travelling people' came to help. Accustomed to a resourceful, *self*-sufficient life, they were not used to doing something useful for others. *We* were not used to so many people on site. The result was wonderful. The aggressively rough language, to defy a society which classified them as outcasts, faded away in the course of the day and all of us greatly enjoyed the atmosphere. For us, progress jumped ahead. For them, the experience of work as gift was inspiring.

A variety of people have worked on the kindergarten. In the earlier, heavy-work stages, the majority have been men, even one who actually enjoyed using the petrol mixer and one who preferred heavy labouring to skilled work. At our 1987 conference, however, which took place entirely on this site, women were in the majority.

For conferences I have always tried to avoid 'wet-trade' work, such as mortaring or plastering, because you can't just down tools to timetable but must finish up the mix and wash up. This time it couldn't be avoided and the timetable, and especially the cook, suffered. All the jobs I would like to offer: dry-stone walling, roof turfing, making gardens, paving play courtyards ... these will have to wait for the future. I have also always tried to have indoor jobs for wet weather. This time there was no indoors and so – of course – it rained.

The fact that we had no roof to shelter under put a lot of pressure on us. Once roof carpentry started we had to push hard to get it waterproof by winter. This coincided with a long wet summer – rain practically every day once we got to the stage when electric tools were indispensable. It was an interesting roof to build – not one where you could do much without a lot of thought. The shapes could only be found by doing things in the right sequence, not by measurement from drawings.

We started by building the ring beams, then we set the disc at the top of the first conical roof on top of a scaffold tower, wedging it to the right level and checking with a plumb line from its centre to a nail in the centre of the floor circle below. Because the ring was made of straight pieces, jointed with steel straps, no two radial rafters were the same length or met the ring at the same level or angle. There was therefore a lot of measuring, bevel-square setting, and arithmetic to ascertain how much the rafter foot should rise if the ring beam cut across the circle. After the cones came a few simple rafters, then ones which seated on sloping beams below, requiring measurements to be taken on one face and transferred to the other – easy to make a mistake with this kind of seating! Then the extension rafters that would shape the fascia. Here was another kind of problem. The fascia would be one of the strongest visual elements – but *we* couldn't see it until we had fixed everything else in position. To make things harder, the timbers we were working with, sized for the weight of the turf roof, were too heavy to easily try out different shapes with.

Shaping the plywood on the cones was another adventure in technique. One of the less practically experienced helpers rapidly developed carpentry skills and evolved a system whereby we would cut one angle on each new sheet to fit against the preceding one, nail it along that edge, then sit on the airborne edge until it was

From the first bricks to the completed roof it was not possible to work with only half one's mind. Although such work requires continual concentration and feeling, curves have such life enhancing properties that even in the worst conditions they were strength-renewing to work on.

(Nant-y-Cwm Steiner kindergarten coming into form)

down enough to clamp with sash-cramps. When sprung to shape it could be marked over the rafter, released and cut with a hand-held circular saw. There was no alternative to using these horrid things, not just because of the pressure of time, but because they could make shallow cuts for bending ply, cut directly over rafters and be used in awkward working positions where it was impossible to align arm and eye.

Both the ring beam, with its thousand 6mm holes 125mm deep, and the plywood depended upon power tools, and both coincided with unremittingly wet weather. We stretched our luck working in the rain, tools covered with a coat during showers or passed through a window into partial shelter beneath the unfinished roof. My confidence in trip switches was reduced by – fortunately mild – electric shocks. I have never got over my dislike of circular saws. They make an appalling noise, and I've met too many finger-short carpenters. This attitude was reinforced when, half-way through the roof, a friend working elsewhere lost a finger.

More than elsewhere, on this site I was aware of a constant risk of accidents. One person seemed to draw accidents to him by his approach of attacking work head on. The risk spread wider, with boards lying around with six-inch nails sticking out of them – all camouflaged in mud and cement. The crane accident I describe elsewhere happened here. It taught me to be particularly cautious of machinery.

On the whole, there hasn't been much machinery on the site. This was a conscious choice, both to preserve the spirit of the surroundings and because of our way of working. Other than a JCB and tractor, a concrete lorry and mechanically offloaded deliveries, we only used a mixer and electric drills – until the circular saws! At the beginning we were offered the loan of three electric mixers, but we couldn't afford to get electricity to the site. So to start with we had to mix by hand. Then we bought an old petrol-engined mixer. It soon became apparent why its last owner had wanted to part with it. Each day's work started, and was coloured by, the half-hour's frustrating effort to start the beast. Eventually its insides deteriorated so much that it started blowing smoke rings, then smoke so thick you could not see through it. Even after hospitalization, and running at its best, the noise and smell were most unpleasant to work with. But none the less I thank it for the endless drudgery it saved us.

Then came the golden day when we were able to lay on electricity and borrow an electric mixer. It started every time, on the mere push of a button, and ran with no smoke, no noise. It made me really appreciate electricity. It was a year after Chernobyl. Now, with three-quarters of the blockwork and all the concreting behind us the work suddenly became much lighter.

Other major sequences besides the electricity supply were done in the wrong order, simply because we could not afford otherwise. Drains, for instance, go through the building walls and stop. It will be disruptive, and in the long run more expensive, to continue them to a septic tank and filter bend, across the workyard and sand and gravel heaps, but we have no alternative. For the same reason we have to carry water from the stream, instead of having a site tap on a supply pipe to the building. This has advantages, however. Working in winter into the dusk, I enjoy hearing the owls hoot as we wash our tools in the tree-overhung darkness by the stream.

Several manufacturers have offered us very generous discounts on their products. Because these need to be delivered as single large orders and not spread over the duration of the job it has been hard to find the money, delaying certain jobs that depended on these materials. Shortage of money after buying these has meant further delays as other things could not be afforded for a while.

Lack of money has not only been the cause of inefficient sequences, needless drudgery and frustrating delays, such as hunting for old nails to re-use, but it is a constant burden to carry in addition to the actual work. I live under continual dread of the order to stop work – although I know that nobody will actually tell me to stop, merely to use up all the materials already on site and not buy any more. This anxiety is much more exhausting than any building work, for its drains one's inner energies.

On the other hand, it is a very renewing site to work on. The invisible life in the woodland is strong and, whenever I can take the time to look around, the river is flowing so close and so peacefully. It was the health-giving characteristics of the place that made me determined that what we would bring by building would not hurt that spirit, but complement it and perhaps strengthen it. So far we have, in general, been able to work in this spirit, keeping the site reasonably tidy, not littering the surroundings with debris, subsoil and trampled mud. It is a 'no-swearing site'. There have been exceptions. One person could swear through the working movements of his body, even if he didn't say anything. He would toss anything unwanted down the hillside – I used to go there to collect half-bricks, blocks, expensive stainless-steel cavity ties and the like.

On the whole, however, the site has been good to us and we good to it.

The surroundings, the building, and we ourselves have noticeably evolved. Selective cutting, favouring deciduous undergrowth has transformed the conifer plantation into mixed, gladed and sun-shafted woodland. The building, even in the stage of blockwork, began the process of continual improvement. The exact focus of windows, turns through portals, floor- and ground-levels, size, shape and location of niches – which, though small, will add much to a room – not to mention the exact shape of the roof, all have evolved on site.

What has evolved? It is not just a matter of little details but of listening to a whole being, something which has never been imagined or frozen on a drawing, something we can only work towards by listening. Only by working in this way can it find its physical form. This, more consciously than on any previous project, is what we are trying to do. The form and place are beginning to grow. The ideas with which we started have determined the way we have gone about the work, but the meeting between ideas and action, between existing place and future intentions, is growing into a living being. A being that is more than either work or ideas.

Steiner kindergarten: a classroom approaching completion. People from at least thirteen nations have worked on this project.

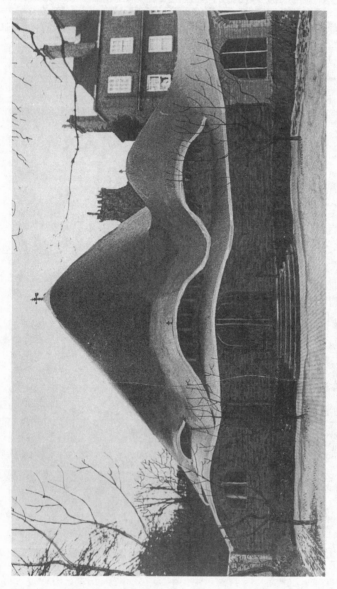

Christian Community Church, London: a photomontage of the proposed building. The inherent stability of curved walls makes them strong even if unprofessionally built. Whereas with paid labour the thought and care necessary to locate and shape non-rectangular forms is expensive, with volunteer work it is free.

Advisory experience: Christian Community Church, London, and Rural Course Centre

IN ADDITION to gift-work projects, the building of which I have been involved in, there are others which I have only designed. The fact that buildings are planned for construction by volunteers has implications for design, building programme, construction information, supervision, site management and the client's social responsibilities. Although neither building I describe has yet been built, issues have emerged even during the planning stage which I consider worth describing.

CHRISTIAN COMMUNITY CHURCH

IN 1978 I was asked by the Christian Community in West London to design a church seating 450 people along with a chapel and other ancillary accommodation, to be built as far as possible by volunteers. My *ad hoc* experience needed to be transposed to another scale and setting, and rationalized and presented in such a way that others could work in their own way in another cultural and social context on another site. To be suitable for its function, the church should have a mobility of form, space and light. What should these sculptural forms be made of? We rapidly discounted reinforced concrete, which would need large quantities of disposable shuttering and involve heavy, dirty and unenjoyable work. Sprayed concrete on wire netting we decided would be even more unpleasant to work with, not to mention 'radar-receiver' problems and a tendency to forms lacking in firmness. Crane-built steelwork would need specialists and also expensive equipment (the crane) on site. Any delays due to low numbers or lack of skill of volunteers doing any preparatory tasks would cause greater expense than were all labour paid. Prefabricated building systems require great accuracy to assemble. Others have adapted or designed such systems for self-builders, but as I myself never manage to put kits together properly, I was not so confident. In any case I felt that the loss of design flexibility – to make non-standard shapes, forms and spaces and to adapt and develop things as one goes along – was too high a price to pay.

Traditional building techniques and materials were, we decided, the most suitable. These in no way imply conventional forms. Indeed, the continuity of local materials and their consequent structural language, with ingredients such as brick arches and flared buttresses, helps quite unfamiliar forms to blend in with those already there. I know many streets with buildings in quite a variety of styles which none the less sit easily together.

Some people consider that unskilled builders are fit only to work on simple forms; anything else needs professional tradesmen. I take the opposite view. In my experience lack of skill is no handicap nor need limit aesthetic aspirations. Non-standard forms demand no greater skill than do standard, but they do require individual attention: thought and feeling. Perhaps every rafter has to be tried in place to see if the curve being set up in the emerging roof looks right. This means that every birdsmouth* may need to be cut individually. There is a simple technique for this. Assuming the rafters seat – as they should – on horizontal surfaces, merely mark up vertical lines from where they touch and mark a horizontal the same height, top and bottom, above what they will sit on. I usually us a marked offcut as a depth gauge.

Since an inexperienced builder is still generally at the stage of giving a degree of individual attention to every piece of work, however repetitive, these extra demands are not contrary to his nature – as they can be with a habitually working, professional

---

* A notch cut in the underside of a sloping rafter so as to form a horizontal surface where it rests upon the member below.

56

tradesperson, whose experience is largely with doing things the *standard,* and therefore unthinking way.

THE QUALITATIVE requirements led the design of the church to a curvilinear form. I could only go so far on paper before I needed to make a model to ascertain whether these forms were actually buildable. I made a 'constructional model', the roof shape built up by cardboard rafters. Just as with the curved roof on Ty-cwrdd Bach or the shaped ceilings at Nant-y-Cwm I discovered that as soon as I ceased to draw, but started to make something, a whole new range of possibilities became apparent. The richness of the metamorphic curves of the roof grew out of this process and at the same time the problem of describing and locating forms for the future builders solved itself. The forms were constructed simply of rafters. The level of the rafter ends on walls, curved in plan, recorded how to form the roof in a way that no contours or other topographic system ever could.

In general, the complex form would require careful attention, but no specialist skills in its construction. A mobile crane would be needed to erect the large semi-portal trusses. Slating anything other than horizontal-eaved rectangles does take some experience. When slating curves you need to know what the slates will or won't let you do. Hence we proposed that a specialist slating foreman should be brought in for this. Subcontractors would also be needed for electrical and mechanical engineering and asphalting. Some other jobs which depend upon craftsmanship, such as windows and doors, or upon speed, such as filling auger holes with concrete as soon as they are dug, would need subcontractors if the volunteers could not meet these require-ments. Most of the work, however, we foresaw as being within volunteer capabilities.

The building would be big – 520 square metres in floor area – and rising over 12 metres in height. Quite a daunting prospect! However, most of its volume would be air, most of the high work would be roof carpentry which could provide its own working access as it rose. The work, however, on a five-days-a-week basis, did not seem so much more than the Steiner school at one day a week.

Right from the outset we knew that there was very little money available. That was why the clients thought in terms of volunteer work. The viability of the project would depend upon its ability to raise funds and, as is so often the case, this depended upon having a design. Even then, it would still have to compete with many other initiatives for limited money.

While I have seen it possible to start building with virtually no money, this project was so much bigger that I felt more cautious. Estimated contract cost was a third of a million pounds. They had £10,000. I felt it irresponsible to advise the client to do too much without much money. I had, however, the very strong experience, albeit limited, that funds flow in relation to visible effort and achievement. Moving from intention to deed unlocks a flow of support and energy hitherto undreamt of.

It seemed unlikely that there would ever, at one time, be enough money to cover the whole project. With the naïve optimism of someone who does not have to sign the cheques, it did not worry me. I proposed that construction – and expenditure – be dealt with in stages. At the commencement of each stage there should be sufficient funds guaranteed to bring it to a state where it would not deteriorate if left open to the weather. This would allow the project to start, and to rise above the ground into public view as fast as possible.

In doing so, it would, we hoped, inspire and encourage both donations and volunteers, even opening up new avenues of support so that this one project would no longer be in competition with others. The supply of gift money available in any society is, after all, not a fixed percentage, but depends upon current social attitudes in general and diverse individuals in particular. The inspiration that glows in society and flames in individuals is a vital factor.

We devised an eleven-stage programme of work accordingly.

Christian Community Church cost programme for construction (February 1980)

Stage 1: Temporary chapel £
Construction, fees (including outstanding fees)   3,150

Stage 2: Completion of working drawings and preliminary preparations
Architect's fees
Engineer's fees
Quantity surveyor's fees
Consultant specialist's fees
Preliminaries
Plant purchase
Total   10,850

Stage 3: Construction to ground level
Excavation and levelling (Contractor)
Fee to neighbour (for access)
Materials to ground level
Total (including, as do all subsequent stages, architectural and
consultant fees for supervision)   12,650

Stage 4: Ground slab and drainage
Materials for slab, drainage and ducts
Total   13,050

Stage 5: Construction of portal-frames
Materials for portals, trusses and structural windows
Foreman for 3 months
Total   12,400

Stage 6: Erection of portal-frames and main roof
Assembly and erection of portals and structural roof
members (contractor)
Walls to main church
Roof asphalt (contractor)
Rafters and felt
Foreman for 4 months
Total   19,450

Stage 7: Construction of vestry and lobby areas
Walls
Rafters and felt
Foreman for 2 months
Total   2,950

Stage 8: Cloister construction
Walls
Rafters and felt

Foreman for 2 months
Total   3,500

Stage 9: Slating
Slates and lead
Foreman for 3 months
Slating foreman for 3 months
Total   5,100

Stage 10: Finishing to chapel
Mechanical services
Electrical services
Glazing to doors and windows
Plasterboarding
Floor finishes
Painting
Foreman for 6 months
Total   20,250

Stage 11: Finishes to main church
Mechanical services
Electrical services
Glazing
Plasterboarding
Plastering
Floor finishes
Painting
Foreman for 8 months
Total   35,000

Such a project would require an exceptional foreman, combining all-round skills, teaching and social abilities with a flexibility of mind, not often found in the building trade. He or she would need to make a long-term commitment to the job. We would be unlikely to find a suitable person without offering a reasonable wage.

Once one person is paid, the economics of voluntary work change. If they have inadequate support, and so cannot work effectively, it is money wasted. Two people can tackle most jobs, one alone cannot. It is thus best to pay two people, both of whom can lead teams of volunteers or, if none come, can work together.

When two people are paid, money starts to flow fast. To make sense of such an arrangement there need to be plenty of volunteers. I have never had a surfeit, but at least I have had some – and from a very sparsely settled area. Knowing that seven million people live in London I had every confidence that there would be plenty. I was told, however, that things aren't like that in London. So we turned our attention overseas.

We now had to think. What could we *offer* which would induce people to travel from the continent to work? London with its rich cultural life could offer a lot. So could Temple Lodge, with its programme of conferences, lectures, workshops and services. The building experience could also give a lot to those who participated in it. We could model something upon the experience of the Nant-y-Cwm workshops, but with a flavour particular to London, giving, we hoped, at least as much to the people who would come as they would to the project.

We began to look into the organizational structures required. We proposed that where school or other groups came to work, that they came for at least two, preferably four weeks, and overlapped the previous group, so that there would always be two groups at a time. The 'old hands' could then help the novices into the job. We suggested that the paid foreman, in order to be able to concentrate on the building work, be supported by an administrative foreman who worked part-time, and that there also be a project group who could take on all the odd jobs such as getting competitive quotations for materials, donations in goods, arrangements for volunteers and so on.

By now we were ready to submit a planning application and did so in June 1980. To our dismay it was refused permission. We immediately appealed, but did not hear the outcome until July 1982. We were successful, but the project lost considerable impetus during the twenty-six months' delay.

The clients started to seek funds sufficient to undertake a major part of the programme. Some of the financial support offered was conditional on the use of contractors, but was inadequate to cover the complete contract. Six years have now elapsed since the outcome of the planning appeal and the project remains in abeyance.

I learned much from working on this project and that is why I record it. However, the proposals which arose from the demands of this new and different situation, were never put to the test. I *think* they were sound ideas, but I do not know.

RURAL COURSE CENTRE

IN 1987 I was asked to design a hostel for residential courses on small-scale agriculture, together with staff accommodation and educational facilities. Visitors would range from city school-children to the handicapped. The project had an important health-giving role – the grounding of modern life on its foundations of the rhythms and processes of nature and agriculture. In line with its therapeutic function we were anxious to imbue the forms with life, avoiding harsh angles and dead-straight lines.

This project is to be built under an Employment Training scheme. What then are the implications for each other of ET and the required aesthetic qualities?

The plan would be complicated to mark out on the ground. The steep slope of the land makes for complications with floor levels and even, indeed, difficulty in taking horizontal measurements. This is a job for only a few people and they need to get it right. After that, the constructional principles are simple: rafters, spanning from ridge beams on posts to external walls make up the varied roof shapes.

There is not much thoughtlessly repetitive work here – blockwork keeps changing direction, many rafters need to be individually cut. MSC schemes have, unfortunately, often been used exploitatively for their free labour, ignoring their educational intentions. ET supersedes MSC and seeks to avoid this. The complications of this project give enhanced opportunities for learning by engaging thinking and feeling over and above manual dexterity and muscle. Indeed the qualitative characteristics of this project will, we hope, add interest, fulfilment and motivation. The living forms and individual qualities of every part of the scheme would be altogether too time-consuming for conventional contractors. The cost penalties (for a charity) would be prohibitive.

In today's economic climate such a building can only be achieved by gift-work, and it needs such a project to transform the compulsion inherent in government employment schemes into gift.

At the time of writing I am still preparing working drawings. Discussions with the probable supervisor make me optimistic that we can achieve our intentions – but this is still a future hope, not yet a past experience!

Playstructure in Liverpool, EASA 1981: working together with the hands overcomes the barriers of different languages.

—— 6 ——

European architectural students assembly: play structure in Liverpool

IN 1981 I was invited to lead an international group of architectural students in the design and building of a play structure on a playground for handicapped children in Liverpool. There were ten of us from seven nations. Tools and materials were provided and we were expected to complete the structure within one week.

I felt particularly acutely that a main objective in my group was to work together, by consensus, not leadership. Moreover, all the preparatory groundwork for the workshop had been undertaken by Nick Domminey, the student who had invited me, and the success of the project was due to him. Consequently, I interpreted my role as that of chairperson – remaining whenever possible in the background and only setting the direction when necessity arose.

However, time was very short and from experience of teaching in schools of architecture I could see a grave risk of our time disappearing into talking, leaving no time to build. We agreed to limit design to one day and devote the rest of the time to building. On this first day I felt it necessary to structure and timetable discussion fairly strictly. We started by airing our thoughts and feelings as to what was missing from these children's environment and what we could provide. At the outset it was apparent that our responsibilities were split between the educational needs of the architectural students and the needs of the handicapped children of the playground. We tried to find a form that fulfilled both needs, but inevitably, in the architectural educational context of the assembly and in the absence of client involvement, the educational aim tended to predominate. We summarized the broad consensus, leaving a number of wilder ideas held only by individuals, and moved on to the next stage when each of us sketched out what we wanted to see. I had hoped to be able to draw qualitative themes out of all the contributions and weave them together into a single whole – not a specific design but enough to imply the initial approach to construction, and to be a starting point of an evolving building.

At this stage, however, two polarized themes emerged. On the one hand, rather elaborate and literal forms of castles, space rockets, the Mersey Bridge and so on – and on the other, enclosures, creating spaces for children within, with varied forms, spaces, textures and sunlight penetration. (We had to guess at orientation as it was raining continually and the playground staff all pointed in different directions when asked where the noonday sun was. Furthermore, it was not clear to us at which hours of the day the children mostly came to play.) Interestingly, the former group, proposing an identifiable symbolic object, were from England and Ireland, the latter group, more concerned with appropriate spaces and qualities, from Europe, Wales and Scotland. With such a range of perceptions, underlying attitudes and individualistic ideas, it was not possible to achieve consensus and it became necessary to vote for a particular sketch. We identified its structural starting point – a telegraph post, from which rafters of varying length and pitch radiated – and started work.

So far we were ahead of schedule. However, we had not appreciated how much of the week would be bitten into by the other conference activities and time-consuming travel across Liverpool. Hence our 'week' became rather pressed. And it rained much of the time. Rain always slows work, but we were compensated by the camaraderie which adversity fostered.

Once work had started, the problems of language communication and of culturally based perceptions and expectations were greatly eased. Working together seemed so much more natural than the self-assertive competitiveness associated with traditional architectural education, and which spreads so easily into the design-on-paper stage.

As the 'building' grew, form, space, light, materials, textures and adjoining landscape all evolved. Ideas arose and potentials became apparent. By mock-up, gesture, or drawing, it was possible to demonstrate and discuss these prior to fixing them in one form or another.

While some other workshops in the EASA assembly degenerated into talking shops or their members lost enthusiasm, our group worked harder and harder. Some people did their own thing – two Finns, for instance, devoted themselves to a Nordic wood-carved dragon's head, another person concentrated on paving and flooring textures – but all was within the context of the whole. For one Jugoslav girl, to handle a saw for the first time in her life was an exhilarating experience. She rapidly became a competent, enthusiastic and exhausted sawyer. For most involved, this had been the first opportunity to design in real form, space, material; to design with the heart and hands as well as the head; to experience design becoming life.

By the end of the week we had begun to establish a good rapport with the other co-workers at the playground who had initially kept apart and regarded us with suspicion, and to crown the week, children appeared in the last day or two to enquire, participate and play.

When time ran out on us, nobody wanted to stop working, but none the less we had completed our project sufficiently and could feel justly proud of our achievement, all within one week, and from an unprepossessing pile of scrap materials.

Looking back over the week, it became clear that through working with our hands, we had overridden any difficulties in communication due to our different languages, and without even thinking of it, had built a strong, mutually supportive group. We had created a 'real building' (albeit of short life) and in doing so, had experienced design as environment; not just on paper, but as a wholeness. This wholeness we felt in ourselves, too. We had engaged thought and aesthetic sensitivity, together with the skill of our hands. And we had given something to the children.

—III—

ESSENTIAL FOUNDATIONS

7

Attitudes to work

LOOKING BACK over some fifteen years of working with unpaid helpers, I realise I have learned a lot by making mistakes – lessons both salutary and inspiring. I can see much, clearly, that fifteen years ago I was not even aware existed. On the other hand, there were things which seemed so clear and easy, so black and white, fifteen (or even five) years ago, which today I find endlessly intricate, ambiguous, fragile – such subtle shades of grey.

I have learned a lot about building of course, but building management and technique few people find inspiring. I have also had to face deeper issues with much wider implications. What, for instance, is the purpose of work: is it just an unfortunate necessity that is hard to evade? What happens when work is organized? When one person employs another either by commissioning work or paying for a product?

Normally, most employment is based on the principle of measured exchange, of buying someone's time and labour. The employee is obliged to work so hard and for so long, the employer to pay so much. Additional effort and time are extras often withheld, resented or paid more for. 'I pay you, so do it' or 'This is what I am paid to do and this only.' Likewise 'If we ask them to take the time to improve the quality, we'll have to pay for it' or 'This job takes so long and that's the time I'm expected to take for it.'

Neither party is free. Enchained by obligations, each restricts what they do or pay to balance that which they receive. This kind of work is governed by the principle of 'gain', and its relationships, procedures and balances are formed accordingly.

When work is *given*, the tasks may be the same but the structure of working relationships and procedures needs to be organized around the gift principle or it just won't work for long. There are practical problems such as lack of skill and unpredictable attendance but these need to be dealt with in new ways. To find appropriate ways of going about things, sooner or later we have to face questions such as:

How can work be inspired without being impractical?
How can it satisfy functional necessities without sacrifices in the aesthetic sphere?
How can it be economic without being exploitative?
How can it be fulfilling to individuals without low productivity?
How can it be democratic without diffusion or loss of direction?
How can it be artistic without the necessary individual involvement becoming egocentric?
How can it be efficient, while retaining the flexibility to develop that which becomes apparent to the listening eye?
How can the necessity of consistency and responsibility be reconciled with the principle of individual freedom – of free gift?

These issues underlie every form of work, but are rarely brought into consciousness. Without the buffer of financial reward, gift-work cannot be sustained if these important issues are not resolved. But more than this, the principle of gift establishes and requires different relationships from those that are found elsewhere.

Even though some people feel one could not afford to live if work were not paid for, some eighteen million people give at least an hour a week: in total, the equivalent of half a million full-time jobs.*

I have worked with those on holiday from full-time jobs, and others with regular part-time work, as well as with self-employed, unemployed, students and retired people. Some had professional backgrounds, others manual. Some had, by my standards, good incomes, others marginal; others again depended on the DHSS.

* Surveys cited in Charles Hardy, *The Future of Work*, Basil Blackwell, 1985.

72

VOLUNTARY WORK: WHO DOES IT AND WHY?

Age	% of sample
18-24	14
25-34	19
35-44	16
45-54	17
55-64	16
65+	18

Sex	
Male	46
Female	54

Marital status	
Single	32
Married	67

Occupational group	
Professional/managerial	18
Intermediate	17
Skilled manual	36
Semi-skilled	16
Unskilled	5
Forces/unclassifiable/housewife	7

Satisfactions	
* I do it because I really enjoy it	51
* The satisfaction of seeing the results	50
* I meet people and make friends	38
It is part of my religious belief	31
* It gives me a sense of achievement	30
Because others are less fortunate than I	30
* It gets me 'out of myself'	26
* I like to feel I'm needed	25
It broadens my experience of life	25
* It gives me a chance to do things I'm good at	23
It makes me feel less selfish	22
Because somebody has to	19
* It gives me a position in the neighbourhood	8

Note: * Commonly stated reasons for job work in other surveys.
Source: Charles Hardy, *The Future of Work,* Blackwell, 1985, p.56
From a survey conducted by the Volunteer Centre in 1981

Why do people give work? In every case I know, the reasons are different. Some find interests or new experiences; some, motivation, appreciation, or, aware of the pressing needs of others, just feel they have to do it. Some find a sense of being contributing members of society that unemployment, redundancy, retirement, home-binding tasks or pressure of financial survival otherwise deny them.

For many people the question is 'What can I volunteer to do? What is there to do that I am capable of?'

As anybody who has any experience of voluntary work knows, anything you can be paid to do, you can also do for nothing! There are very few specialized skills into which the working world is divided that cannot be picked up sufficiently by non-specialists. After all, any new job is likely to be at least a little different from the last. Volunteers may not do the job very fast nor to a high standard, but the need is usually only for an adequate standard. When undergraduates (whatever we may think of their political leanings) ran essential services during the 1926 general strike, the results were not disastrous. Sometimes the situation requires people to learn new skills on the job, sometimes it is more a matter of adapting one's experience to make it relevant to the need. This is particularly so for people made redundant who carry with them skills and experience of great value to society, but for which no one will pay them. Everywhere there are jobs which need to be done, but which fall outside the sphere of economic necessity and therefore there is no money to pay people to do them.

When you make a list of your practical, managerial and domestic skills and experience, your hobbies, interests, concerns, and energy, and the personal problems you have had to work through, there is no one who does not have something to offer. And whether in organized projects or personal initiatives of neighbourliness or environmental care, the opportunities to contribute are there. It is a cruel irony of modern civilization that the most important work cannot be paid for. It, and with it the quality of the society we live in, depends upon gift.

There is also the question of who can afford to support such work. In reality, donation is not a function of riches but of priorities – of values. People who give are in their own way working to subsidize *work*. In this case, however, they are subsidizing work by the hands of others. As gift, which is the greater: £1,000 from a company, or £1 from a social security claimant; half an hour from a mother of a large family or a sustained month of work? The value in material terms is different, but to think about worth is a demeaning way to think about people. I need constantly to remind myself that once deed becomes gift, it is all gift, and not to be weighed, judged or compared.

It is said that times are hard; nobody can afford to give like people used to. Yet at the end of the Second World War Britain was effectively bankrupt. In those depressed economic circumstances, the ideal of National Health was born. The ability to give depends more upon concern for others than upon the availability of material surpluses. Surpluses these days are destroyed by the million ton, grain burnt in the USA, fish sprayed with diesel in the EEC.

As I have learnt through experience, gifts arrive if one can inspire the giver. After a short magazine article on the founding of Nant-y-Cwm Steiner School, I was surprised to receive a supportive letter and a £10 note from an anonymous donor. It reminded me of the early days of building when, arriving to work one morning, we found a pound note pushed under the door into the building rubble. In neither case had we asked for money. The donors had been inspired by our work; we were reinvigorated by their gifts. Donor and recipient had been enriched by each other.

The relationship of giver and recipient is established by the medium of money. How we view money affects what we do with it, and indeed what it does to us.

Money can be viewed as the lowest common denominator of wealth, the substance for which wealth of various forms, art treasures, landscape heritage, for example, can be exchanged. Real wealth is that which enriches the spirit. It is at the heart of civilization. But money can threaten civilization. It is an outer form only, but with a magnetic power that has led many to regard it as godlike – or demonic.

When I think of money in society, I think of the nutrient cycle in nature, where organic matter cycles out of life into chemical substance and thence back into life again. In this sense, money is the lifeless chemical substance. Will it stay lifeless and draw to itself more and more ('the easiest way to make a million is to have £999,000') or will it enable life? Can it support spiritual inspiration that wants to work in the material earth?

In inspiring donors to set money into life, the recipient achieves more than he or she receives. If nothing else, gift-work must bring us into a revised awareness of the relationships between money and wealth – matter and spirit, between donor and recipient – supporter and steward of an ideal.

Charitable work gives the opportunity for the fruits of dead work to fertilize new work, as vigorous as it is inspired by ideals. In giving money, which has condensed out of old wealth, we can enable new wealth to come into being. Here, I think of a friend who sold a picture collection to raise money to take an education, eventually to teach.

JUST AS we need to revise our approach to money, so too do we need to reconsider our attitude to work. We can look at work as something by which we live or *for* which we live; as taking or giving; for its material ends or its spiritual purpose. Over the centuries attitudes to work have lost sight of its spiritual function. Once, the buffalo hunt was as much in reverence of the buffalo, in praise of the Great Spirit, as it was to provide food, clothing and tools. The earth was first ploughed – opened to the heavens – as a sacred act; only later did the plough become a tool to initiate biological processes. Work as a concept condensed into a separate, isolated element, set apart from other aspects of daily life.

In recent times work has become merely an inevitable and inescapable habit, often loveless and devoid of pleasure, characterized as the protestant work ethic. No longer is it common to spend one's life visibly working for the needs of the community, visibly supported by that community in the form of earnings. Now it is normal to work because one has to, because one needs the money. No longer do the carpet makers dance the weave.

Dozens of small children are assiduously holding slender black threads looped around their tiny fingertips. The threads stretch in an intricate pattern from one cousin to another brother. In one corner a wrinkled old grandmother works at a spinning wheel, and a small girl is waiting to take her freshly spun yarn to the dye market. The men are performing an ancient dance, the dance of the rug makers; by gliding in and out among the little children while knotting the warp-threads. They move according to a cadence which is sung by the women, who are sitting around the walls paying out measured lengths of coloured wool, in a ritual drawn up unpolluted from the deep well of time.

Each district, each family has its own special song, and this gives each rug its unique design. With each change in rhythm comes a change in colour; a new harmony makes a new pattern.

David Saltman: 'Decoding Arabic design', *Shelter*, Shelter publications, USA, 1973, p.130.

Through work one enters in a special way into relationship with matter. Behind every conscious action lies an idea, but to bring that idea into material reality requires work. And work brings us face to face with obstacles, time and effort. I may have a vision of a beautiful building but all I find are concrete blocks, dead-grey, heavy, hard. It is the constructive and life-filled meeting of the inspiration and the obstacles which create the artistic physical result. And that takes, and gives meaning to, time and effort.

In India Beauty is elevated into godhead, interchangeable with divinity, in fact worshipped as Beauty, . . . the source of life. It is therefore inevitable that this reaching out to beauty and its manifold manifestations has been a constant factor in the life of the people, in rituals and ceremonials, festivals and celebrations . . .

Their abode of dwelling had to be something more than mere four walls and a ceiling. Every article used had to be elegant: lovely to look at, elevating to live with. Beauty in this concept is, therefore, not divorced from life as lived every day. The commonest object is endowed with grace and colour. Every act and function human beings perform must be touched by this magic wand. Beginning with the harsh, severe walls of the primitive dwellings to baskets and pottery, from clothes to objects of worship, all was involved with this grand spirit.

This Beauty is not confined to hanging a frame on a wall, an image in a niche. The objects are not meant to be kept in glass cases as status symbols. For anything superfluous in the traditional world would be an object without a significance, regarded more as a sign of human vanity, not creativity . . .

Utility is the necessary part in the completeness of life. Through aesthetics in utility, beauty is brought into our intimate life . . .

Just as human beings learnt that health can be maintained only by respecting the laws of nature, they also learnt that where their handiwork failed to acquire the excellence which is being yearned for and the eye sought, they knew there was a lack of co-ordination between their concept, the material they used and the method they employed. This fusion was the fundamental precondition for quality – the mastercraftsmanship.

Kamaladevi Chattopadhyay, 'Craft traditions, a continual renewal', in *Resurgence*, 123, July/August 1987.

It is not so many generations ago that peasants themselves made everything they needed, from houses to furniture and utensils. In other parts of the world it is still the case. The artefacts were never ugly. Many people find them beautiful, more satisfying to look at, feel, hold, use, than the modern equivalents. At the heart of this, stands the attitude with which things are made. To so-called primitive people, a soup-ladle was not just a utilitarian tool; it was also bound up in the act of ladling soup, the whole process part of giving thanks to God for daily food. Every element in daily life was in a way sacred, every repeated deed, like taking the cows to pasture, a ritual. There are still legacies like the belled Swiss cows, the music of which far exceeds any utilitarian requirements. To such people, utility was inseparable from beauty. Not to make something beautiful would have been unthinkable, sacrilegious. Modern design tends to *apply* beauty to utilitarian form. It *follows*, is *separable* and is often added to imprint personalization or enhance value. Indeed, even in my practice, I am from time to time asked to justify aesthetic decisions on the basis of enhanced value.

Usefulness to others is readily visible, but there is another level at which one has to ask what work is for. Can it be a vehicle for deeply held values to take substance – adding, quite independent of monetary value, to the real wealth of the world? How different from work as imposed drudgery!

What, other than bitterness, has ever been the personal fruit of work that one resents? Occasionally I have had to do things I thought unnecessary, unmeaningful. I regarded such work as a good exercise for the will, but I could never keep at such jobs for months or years. If those jobs also contribute to harm and ugliness in the world, then to continue at them is unhealthy for society and stores up the seeds of illness in the individual.

But if, out of an ideal, deep within oneself, one *wants* to do a job, that job – whatever it is – is nourishing. It need not even be the sort of work you might naturally choose. When, fundraising for a charity, I worked under rain-lashed canvas nineteen hours a day at double speed, cooking, serving, washing up in an atmosphere of mud, crowds, noise and discomfort, it was nevertheless an exhilarating experience. When customers came to the back of our food tent and told us that our queues were the longest, our food, service, atmosphere the best on the nightmarishly vast festival site I could answer from the bottom of my heart: 'I couldn't do this for money.' Were the work governed by pecuniary arrangements, I am sure I would have already come to blows with my employers. It was exhausting, but none the less more nourishing than stressful. Why? Because our work was *giving* service – and was appreciated – only *secondarily* it was for money, and that also for the benefit of *others*. Our work was inspired by an ideal, and this was sensed by our customers. Instead of the aggressive scenes that all the other caterers experienced, we received unsolicited praise which gave us new strength when we were wilting with exhaustion. Approaching work as gift, we received in return nourishment.

Limited as it is, my gift-work experience outside building confirms my belief that this book can be read as an illustrative example, a way of going about things relevant in other spheres. The principles and their implications for how work is approached can be transformed to any work, whether supported financially or not. Any work and all work.

———— 8 ————

Why work anyway?

FOR MANY people, work is something they have to do to buy the
necessities of life or compensations for their soul-destroying work.
But work can be much more important to us than that. In our
daydreams we can change the world without getting out of bed.
We can also work all our lives without recognizing anything
spiritual in it. But when our dreams become united with deed, we
have a star to inspire our work, and practical opportunities to give
concrete substance to that inspiration. For all the assumptions
made about the profit motive fuelling work enthusiasm, research
sets identity, friendship, status, the sense of creating something,
achieving and contributing, well above monetary reward – which
came twenty-fourth on one survey list.*

Work is part of our biography. In it we find opportunity to unfold,
develop and take new directions, to grow. Satisfying work allows
us to contribute meaningfully and beneficially to society. A
working economic relationship is one of mutual benefit: one party
gives and one receives. Both are enriched by the process.

Of course it is possible to work for oneself – to be self-sufficient,
individually or as a community. But for *whom* am I really working?
If it is just to fulfil my daily needs, the only differences from
employed work are that circumstance not the employer curtails
my freedoms, there is no cash exchange, and the hours are longer.

* Cited by Charles Hardy in *The Future of Work*, Blackwell, 1985.

But it can also be work to experience and express the beauty and sacredness of creation. Many self-sufficient small-holders experience this – that their work is not just to support the physical mechanics of life but to participate in something that lives in the realm of ideals.

Over the centuries, work has become a compartmentalized concept within the whole of life. How could the primitive tribesman distinguish which was work and which leisure when he hunted or danced. Both were sacred and necessary. And how indeed can we say that a woman carrying water a mile from a well is not working, while an overseer, patrolling the factory floor, is? The concept and definition of work have become tied to monetary criteria. Housework, gift-work and study are excluded from what is now the widely held meaning of the word, yet they all involve effort, resistance and change. Work is work whether or not it is productive or income-earning.

> The food grown as well as the food cooked or served, the carpet woven as well as the carpet sold or delivered, the house built as well as the house cleaned, create the space in which we as human beings can live together. The difference between the so-called productive work and the so-called unproductive work is that the first is more apparent to our physical senses. The fact that a house is there appears more weighty to the life of the people living in it than the fact that it is kept clean and tidy. And yet everyone who has lived for some time in a community will know how much the hardly perceptible fact of the daily cleaning and tidying contributes to the possibility of living together in harmony.
>
> Baruch Urieli, 'What is the fruit of work?', *The Threshing Floor*, Floris Books, Edinburgh, January/February 1987.

Unfortunately, work is all too often tied to income. If we give ourselves through our work, we are not selling ourselves. The taste is too sour when someone tries to buy me, yet I ask for money for work. Work and sustenance (usually in the form of money) are interlinked in many ways but the more distinct they can be made, the freer is our work to allow us to give of ourselves. If we can picture an ideal community in which work flows through the economic process in a cycle of giving and receiving, we would also see sustenance flowing in a similar manner – not as payment but as work and need. There is nothing impractical about gift-flow economic structures. Many schools, clinics, homes for the handicapped and other initiatives seeking to apply the social insights of Rudolf Steiner, neither charge fixed fees nor pay fixed salaries. They neither buy nor sell work, but ask for freely given contributions out of which to fulfil the varied needs of the individuals working there – who remain free in their assessment of their own needs and ability to give. This relationship of freedom allows both work and sustenance to live within the streams of gift. How different from the static equation of reward taken in exchange for effort grudged, for gift is alive!

Gift enriches both the recipient and the giver. The more one gives the more one grows. Through gift people give to others new inspiration and courage. Work has gift inherent within it, waiting only to be recognized and consciously acted upon.

As the moulder-machine operator adjusts the machine to its repetitive production of plastic toy parts is he or she contributing to the wealth of the world – by helping children to be happy perhaps? *I* have strong opinions about plastic toys, but the operator must make his or her own *free* assessment – is the job for others' benefit or for the wage? The first is giving – much more so than I can ever do, for I do not face monotony day after day in a lifeless artificially lit environment! The other is a waste of the gift of life.

We die different from how we were born. The road of life has brought us many experiences, struggles, joys, sorrows and traumas which we have had to pass through, and which allow us to shape our inner development. The many long hours of work are part of this process. Ten years ago I tried to project the future. What would I be doing in my work, how living, in middle age? How would I be? The developments, the unexpected avenues, turnings, even dead ends, which led to new light beyond, were all quite beyond my imagination. Destiny brought me to face problems in myself that I did not even know existed! Looking back, everything falls into a pattern which led forward. Looking forward I can only see as far as the limit of my present consciousness, and imaginative pictures which are no more than variations of the present. Thinking about things could – and did – bring me no further. Living them through active deed did.

The bringing together, with effort, of inspiration and matter, the meeting of need with service through gift, brings a stream of health into the world, opening at the same time the gates of inner development. Through work, we can leave the world richer than we found it, ourselves richer than we entered it – spiritually richer, for as my old neighbours say of money 'You can't take it with you when you go.'

Without work and good will there would be no society. Yet powerful forces are undermining both. Not only is the exchange of human time, energy and gift treated as exchangeable with financial capital, but the purpose is often seen to be production; the desire for products has begun to rule our society and values. Many people think of work as a necessity for income, regardless of whether it makes any contribution to the world we live in.

In one sense the conflicting perceptions which bedevil labour relations in the capitalist developed world are but two sides of the same coin. These polarized forces dominate and debase the political scene. Society needs a renewed vision of work – work as practical good will.

Through work we establish relationships with the world around us. As our natural surroundings have become progressively de-deified, no longer valued as God's creation, they have been seen in more and more material terms, from the perspective of their potential for use and exploitation. Things have always been taken from the earth and living world, but when the killing lost its reverence and became slaughter, and the materials were not raised to become beautiful as well as essentially useful, the way was open to thoughtless, heartless and often needless despoliation.

Work with these objectives is a taking process. Global environmental problems are an inevitable result of such an unbalanced one-way approach. Similarly, excessive emphasis on material production leads to the use of people merely for their achievement. Their productivity is maximized by atomistic specialization, but at the expense of human balance.

You can buy a window or you can make one. As long as you ensure that water will run out of joints etc, that glass will not need to be bent and that the finished object will not distort, it will be workable. But why have all the expense to buy one, or the effort to make one, if it is not beautiful? Windows (or anything else) do not stand on their own; we experience them in relationship with their surroundings. Once you think of how the window will speak to the wall-shape and around it, it is no longer possible to regard aesthetic considerations as separated from practical ones. In such a way, we can re-establish, through conscious activity, the artistic approach to craftsmanship that in previous ages was intuitive.

Many jobs are disproportionately one-sided – employing only hands or head. In some, employees are chosen solely on the basis of dexterity or intelligence tests. Society pays the price of this process of dehumanizing by dismembering, as frustration and boredom find release in violence.

We consider archaic the traditional shipwrights who could not bring themselves to build boats that were not beautiful, did not sail well or were not, from invisible structure to painted finish, well made. Yet I have talked with old bricklayers who miss the skill, demands on their thinking, and aesthetic delight of the arched, curved, bonded and patterned brickwork they were brought up on. Nowadays, as such people say, a supertanker is a floating box designed and built as cheaply as will last till the Middle Eastern oil runs out; a bricklayer is hired on the basis of how many thousand bricks he can lay in a day. But if he was concerned about the aesthetic consequences of his work, how long could he keep it up? I certainly couldn't work on the average building site with its feelingless end products for very long.

Cultivation of aesthetic sensitivity helps to raise our feelings from the desire plane to the artistic. The absence of the artistic from daily work which would infuse the work of head and hands with sensitive feeling, has therefore profound implications. Art transcends the physical and cerebral to nurture the higher self. Deeds based on artistic values transform, indeed redeem, matter, reversing the materializing tendency of our age.

The sculptor does not think about how most easily to do something, but how to enhance and refine that which speaks to the soul. The economy which concerns the container designer is monetary. How can his container be either cheaper or more attractive to a purchaser by its durability, performance or style? He strives for it to be cheaper *and* more attractive. Then he can sell more. The artist strives for his work to touch something universal in the soul. The aim for one is pecuniary, the other spiritual.

Art is not just the prerogative of professional or even amateur artists. In the sense of making every common deed meaningful and full of beauty, it has a vital place in all aspects of daily life. It is that which distinguishes reverent from loveless work. Sacred work can only be undertaken in the spirit of gift. Whether paid or not, artist and priest give themselves to their work. Whether paid or not, neither work for money. Although it may provide the necessary sustenance, when money becomes bonded to work, the elements of gift and reverence have gone. While all work can be undertaken in this way, gift-work forms a vital demonstrative role, and a no less important experimental one as the consequences of new ways of doing things are worked out.

9

Gift-work

IF WE take as a definition of art that which raises the material towards the spiritual, there is no aspect of everyday life that cannot be raised to become an artistic experience. Every action can become a sacrament.

When designing a church I initially regarded the vestry, where the vestments are kept and changed into, as little more than a storeroom. In discussions with the priest, however, I came to understand that here they are ironed, hung, laid out. These daily home tasks are here a conscious sacrament. In the same way even the most lowly tasks of everyday life can be raised – even wiping a child's bottom! Try it as a sacrament – what a different experience from the same action performed just as a necessary chore!

Nowadays the aesthetic, caring craftsmanship, meditative and wonder-giving elements have been reduced, if not completely eliminated from so many activities. Compare the wonder, rhythmical work and sensory richness of baking bread with the experience of work in an automated bakery. Or the miracle of the sunset which used to set in motion the transition from outdoor to indoor space and society with the lighting of the lamps and withdrawal to the hearth. The norm today is electric lighting and constant temperatures.

Activities which today we regard as purely functional, be they even as mundane as heating, harvesting, weaving, were, to the ancients, religious activities. In medieval times, the craft institutions had reverence for work, responsibility to the community and a moral code. Daily work was a spiritual task. Throughout the pre-industrial era, work and wisdom went hand-in-hand. The apprentice absorbed both as an inseparable whole.

Gradually the effects of the universities, the attitude of specialism, of separation of knowledge from active physical experience, began to develop into scientific research, accompanied by the emergence of an intelligentsia. The arts, themselves becoming more specialized, were increasingly available only to this educated élite until they became so isolated that they could be considered as playthings of those people who are generally separated from the actual work of providing usable goods. And so they have remained. Art and utility became so separately compartmentalized that sculptures and murals could be incongruously applied to the walls of starkly utilitarian buildings.

Even if, in full consciousness, we seek to imbue our daily, often humdrum and habitually graceless actions with an artistic quality how are we to do it? Although it can also be temptingly misleading, it is worth looking at the way things were traditionally done for they were almost instinctively artistic.

To build a house the pre-industrial peasant took of the earth, stones, trees and straw around him. With these materials, together with his brothers, cousins, uncles and friends, he constructed a building, the stereotyped design of which was imprinted over many generations with but slow evolution into their image of 'house'. In the process, the products of his daily work and surroundings – his complete experienced world – were combined and raised to make a home. The surroundings, social community and archetypal idea were combined into one whole – no wonder we find such buildings attractive.

The limits of our surroundings, bonds of community and expectations of comfort are different today. Today our building materials are not local, our constructional methods not bound to timeless structural principles – we can use invisible reinforcement in concrete for instance, to make shapes which without it would fall down. Our image of a particular building type – such as a home – is at worst plucked from, at best synthesized out of, a catalogue of ideas.

Where, in the past, habit determined *how* things were done, today our actions involve an element of choice. Things made these days may often be less attractive than the products of the pre-design era, but we can do something our predecessors could not. By committing ourselves consciously to the ideal of beauty, our deed can be a conscious gift, whereas theirs was a habit.

Some look nostalgically to the old ways as so much better than today, while others point out the cruel hardships and inequities that many had to endure without the buffers of modern technology, medicine and social support, but the real issue is that they were different. In no way can we return to that consciousness. At best we can only ape its outer forms.

What we can do, however, is learn by the past so as consciously to rediscover its virtues and overcome its abuses. Why, for instance, is not our modern work artistic and sensorily rich? It can be. Steel must be painted to protect it from rust and there is no utilitarian or significant economic reason why it should not be painted beautifully. Nowadays even the colour of paint can have a functional purpose such as solar reflectivity. If we commit ourselves sufficiently to the intention, something akin to the restful and poetic qualities of Mediterranean villages can be achieved. If not, the quality of white chippings on a flat roof is more likely.

Lack of consciousness and sensitivity is not the only reason the environment is so debased. Many things, not only cardboard structured doors or pine-look furniture, are meant to look attractive but without costing too much. The wider the discrepancy between apparent and actual cost in time, effort and material quality – all of which the manufacturer can cost – the larger the profit.

In building nowadays, craftsmanship has largely given way to speed. Opportunities for tradespeople to contribute over and above their contractual obligations, let alone artistically, have almost disappeared. Efficient management strategies separate trade from trade and rob workers of a total view of the project on which they work.

Inspiration is the essential foundation of reverent, artistic, given, work. Yet it is not always so easy to find inspiration in a world outwardly fuelled by profit. Some of us are lucky enough to work for charities, producer-consumer organizations or businesses whose aim is to bring benefit to humankind or the environment.

Coleg Elidyr schemes for handicapped young people

1. Apprenticeship – A pathway to human dignity

In an instructive environment with an undercurrent of motivation, in a setting of peace, immature young people not yet able to face an adult world and a competitive work-situation, can learn to apply themselves and achieve a discipline and skill.

Small teams – dedicated workmasters, a progressive learning situation. From rough and ready to skilled work, to responsibility and independence. A skill-curriculum, adapted to the individual need makes it possible to complete the learning process at one's own speed and in one's own good time.

Choice of apprenticeships:

Farming, gardening, herb nursery

forestry

pottery, weaving, printing

general building

carpentry, bricklaying, plumbing

guesthouse management, café

food preserving and processing

2. Journeyman scheme – A pathway to meaningful integration and living

Few openings can be found for the 21-year-olds who are making their first tentative steps into a work situation in society and the newfound confidence and courage may falter.

The journeyman scheme is a tentative integration guided and monitored to consolidate their know-how in an objective and production-conscious setting but without the stress of competition.

New work-experience and responsibility in ventures near by or further afield, even abroad, will provide new dimensions to the working and social life.

A journeyman will learn that though there be a shortage of jobs, there is never a shortage of work and that there are always people or situations needing help and selfless application.

Coleg Elidyr Camphill Centre, Journeyman Scheme Prospectus.

While it is possible to *give* work under any circumstances, many jobs do not provide a supportive environment. Nor do they, in return, nourish the worker, for so many jobs are one-sided and fragmentary and separate the workers' actions from the users' needs. What a difference it would make if the assembly line worker travelled the line with his growing product, from start to finish, and personally handed it over to the customer! What a difference if he or she could be just a little involved in the workings of the complete system – unlike the mechanic who got our cement-mixer running beautifully, but didn't worry that the chain didn't fit the drive cogs! What a difference if the worker could see the aesthetic implications of his or her work.

The modern building process has become modelled on the high productivity factory. Some sites even have their own on-site factories for producing components. Building is called an industry. Much work is serial, atomistic, skilled but without understanding, and with no care for appearance other than the standardized standard. Such work is not fulfilling. Its onesidedness does not develop the human being. It may be profitable, but it is not health-giving.

Volunteer and self-building provide opportunities more or less denied under the contract system, for work at all stages from design to maintenance (but especially during building) to engage mind and heart as well as hand. Elsewhere, artistic input can rarely be afforded. When it is, it is provided by specialists. The contractor's economic viability depends on tradesmen who know what to do so well that they don't need to think. When they meet the unexpected and have to work out what to do, down goes their productivity and his profit becomes the threat of bankruptcy.

In given work, beyond the opportunity for hand-finished, artistic and well-thought-out work, there are further benefits for the recipient (client). The most obvious of course is the provision of buildings at reduced cost. There is also the infusion of quality, not just visibly but also invisibly yet none the less perceptibly. Near Findhorn in Scotland, I have had the experience of divining a house of which no visible remains now exist. My teacher with divining rods, and I with my hands, found the location and size of the stone walls, partitions, doors, fireplace and roof, and where

What adults call work can to children be fun. How can we arrange daily work so that it is something we want to, rather than have to, do?

people and animals lived. He was further able to date the building and its destruction – 1747, the year after the battle of Culloden, and record fear as the principal emotion of the people who lived there. It added a new dimension to my understanding of building responsibilities.

The attitude which is imprinted into the results of working is so important, and can be so varied. Children often play at work. Indeed all imitative play can be seen as adults' work from which necessity has been removed. In play, children do things for the sake of doing them. Adults, working, do things for the results. I remember as a student many things I did were for the sake of doing them, as much if not more than for the results. Hitch-hiking for the pleasure of travelling, intensified by diverse company and experiences; working through the night for the experience, which a little forethought would have rendered quite unnecessary.

Something of this quality of enjoying every experience – not all of which are fun, for thumbing for seven hours on a motorway slip road in the rain isn't really fun – can be part of everything that one does. It certainly takes an effort to turn unattractive work into a consciously appreciated experience, but in most work there are more possibilities to do things than are taken up. I have worked in the rain both with complainers and with those who could make every slippery mishap into a joke. All this is imprinted into the quality of the work we produce, be it motor repairing or business communication.

The bond of obligation between employer and employed can squeeze pleasure out of work. Employed, I am obliged to do unpleasant jobs, because I am paid to take orders. These jobs must be done, but as a gift-worker, I have to summon up my own free will to overcome the unpleasantness of the task, whether it be digging out a used septic tank filter bed or hours of paper work in beautiful spring weather. If I am being paid, I wonder if it is worth it. If I am giving my work, it is gift, not exchange – anyway many people have much worse tasks as a matter of daily routine.

If you are paid to work, your output and the employer's output of money are measured against one another. In any commercial business the employer needs a profit to survive. Other than in co-operatives it is rare that any profit is meaningfully distributed amongst its members, or to charities. Elsewhere it goes to the employer or the shareholders. Even if it is no one's aim, the conventional employment framework cannot avoid exploitation.

When a new building company was formed between friends, only one member was prepared to take any long-term financial responsibility, so it became not a co-operative, but an employer-employee firm. The employer naturally sought to make enough to protect himself. Within months, trusting friendship had become eroded by a feeling on the one side of exploitation, and on the other that the workforce was no longer really seriously concerned with the efficiency of their productivity or the quality of their work. 'He's doing all right out of this' and 'it'll do' had started to enter into what had started as a commitment to co-operative quality work. I had the same problem when I took on an apprentice. Some of the risks could be eliminated by asking clients to pay him directly. But there were still office and time overheads and – hardest of all to cost – the risk of follow-up time long after any job had finished. I charge an overhead factor on this, but at least I do not need such a large defensive margin as I would if I carried all the risks.

It feels bad to be exploited. It is unhealthy and the resentments it breeds sour the social relationships in the workforce – the chargehand for instance becomes the employer's agent. It lowers the quality, visible or invisible, of the work created. The lovingly home-cooked meal is replaced by the dollops of canteen-stewed cabbage – to a greater or lesser extent in every kind of job. Work given out of freedom removes this burden of compulsion and makes it easier to fill work with willing enjoyment in place of sour resentment. Out of the bond of working together on this basis, the social links of community start to form.

MANY PEOPLE, friends, friends of friends, former students and friends of students came to help us build our first house. The place became a meeting point of many faces, many memories. When I look back it becomes clear to me that the guiding motive was to create a work of art not a tradable commodity.

When I look back over the building of 'my own' houses, I realize that had I wished to own more or enhance property value or personal prestige I should have gone about it in quite a different way. I wanted to create a home for my family, but more than that I wanted to make something beautiful. It was that ideal of the creation of beauty that sustained me through the long and arduous building process. A self-building client of mine, not a sculptor, told me that for him, building his house was building a sculpture to live in. I've stayed in his house, and it is.

I have worked on my own, in heavy, sticky mud. It was quite soul-destroying. I have also worked, as part of a team, in mud so bad that we sunk in, wellington-deep. To take a step I would need a pick-axe to lever out my foot. On my own I would just have given up, but, as part of a team, humour and determination became the keynotes. In a curious way, we enjoyed it.

It is similar with situations that border on the dangerous. In awkward positions, on a rain-slippery roof, struggling with a wind-torn tarpaulin to cover unfinished work – on one's own the task would be desperate, dangerous, impossible. With someone to work with, the feeling is more one of determination, the risks can be controlled, the task achieved.

There are always dull jobs that have to be done. Moving materials is pure drudgery for one person. A team does the work faster, effecting rapid transformations. Often its individual members can be brought into one social whole by the rhythm which unifies their work. What a difference there is between individually carrying blocks across mud, obstructions and change of level, and passing them from one to another in a smoothly flowing chain of movement.

Traditionally music and song accompanied heavy work. The 'navigators'* who built Britain's railways and canals would stand around a stake, their sledge-hammers rising and falling in turn around the circle, to a sung rhythm. Anchors were raised to the rhythm of the sea-shanty, a fiddler sitting on the capstan around which the seamen toiled. Within living memory, milkmaids in some parts of Britain sang to their cows to the rhythm of their hands on the udder.

* Hence 'navvies'.

Give up thy milk to her who calls
across the low green fields of heaven
above the hills of paradise
gain and save thee and keep thee for ever.

(St. Bride's milking song – attributed to Fiona Mcleod).

Even though many people feel shy about singing in public there is scope to develop rhythmical forms wherever things are moved – passing chains, throwing-stacking pairs, up-and-down-scaffold handling of goods. All these are a pleasure to take part in which cannot be said for carrying a hod of bricks up a ladder!

We can try to arrange work so that individuals feel important members of a productive team; where their understanding and aesthetic sensitivities are involved; where the work is a pleasure to undertake and gives a good feeling to take home with one; where one wants to give oneself rather than merely fulfilling contractual obligations. Such work nurtures the whole human being. How different from that that many people have to suffer! Opportunities for fulfilment and nourishment are so often curtailed, concealed or absent in conventional work. In gift-work situations they are more accessible. A special responsibility falls therefore on the organizers of any gift-work project to bring to consciousness these themes and their possibilities. In this way, the consequences of gift-work are not limited to the works of any particular charity, but are the seeds of change.

10

Building

FOR MANY ventures, buildings are essential for their birth, development or growth. Yet it is an unfortunate fact that buildings are *very* expensive; expensive to build, to adapt, or to purchase.

Many charities can run effectively and vigorously on next to nothing – until they come to the need for a building! You can do a lot with £50 a week. Yet this could buy only one or two square feet of building floor area. The economic scale is now magnified perhaps a hundredfold or more. Charitable initiatives serve as channels for the desire that many people have to serve others, or to make a meaningful contribution to society that unemployment denies them. But at the point at which an initiative needs a building, the money required is just not available.

Just as for the would-be home owner, it is at this point that thoughts turn to self-building. Initiatives which would otherwise be stillborn are now possible. Buildings can grow at a speed appropriate to the enthusiasm, support and cash-flow of the group. Even if it does have the money, this flexibility is greatly restricted if one buys building work. The substance of the building itself and the effort that goes into it can become an inspiring symbol for the project which it serves. It guarantees that in the area of heaviest expenditure, money given will go much further than it would elsewhere, so that the real work of the charity is not disproportionately drained financially.

Volunteer building has many advantages: economy, focus on

vigorous activity instead of the institutionalism which buildings can bring, community building, personal growth for those involved and opportunities for artistic work. But it is not without problems! Economy has its price in energy and duration; gift-work in unreliability. Matching building with cash trickle is wearing in the extreme, as are conflicting demands for space by users and builders, both *giving* work, without contractual demarcation. Indeed there is a multitude of disadvantages as people have pointed out to me – and I myself have discovered enough! But while the disadvantages are identifiable and visible, the advantages are less so, but greatly outweigh them. Even though, at the time, the problems seem so great, when I look back at mature projects, I am amazed at what has been achieved.

The problems however must not be underrated, for, without care and a proper approach, volunteer building can be a disastrous, expensive and embarrassing failure. With care, commitment, sensitivity and forethought, it can set in motion much more than merely providing a roof to keep off the rain.

While this outer need of buildings to house activities is so apparent – and is invariably the starting point – the necessary new approach to work, with its new structure of relationships, opens up so many possibilities for work, society and art to develop. What outwardly appears so important can be seen as secondary to these developments with their profound implications for society. The desire for an attainable, cheap and adequate building is really only the starting point.

OFFICE AND manual work tire in quite different ways. Manual work, physical exercise, sport, bring the sort of bodily exhaustion that helps you to sleep. Office work, mental work, winds you up.

Many people find limited periods of physical work an invigorating relief from mental work. Some feel that it should take up a part of every day to balance body against brain and action against thought. Monasteries, Maoist communes and many other kinds of communities do this. It is widely accepted that daily exercise is healthy, but manual work is not the same as exercise and it does not have the same effects. For health, exercise should be measured, balanced and physically symmetrical. Manual work, however, is usually scaled by the requirement of the job rather than the person, often involves principally only one part of the body and is frequently asymmetrical, localizing strain, particularly on the spine.

Overstrain causes injury. Manual work, particularly for those who are unfit, out of practice or not physically adequate to the demands placed upon them must be measured with care. Yet when you are paid to work, you are paid *to work*. When, however, you volunteer to work, what you are prepared to give, not the terms of the job, determines the measure of your work. As many people don't realize, or won't admit, their limitations, responsibility rests on the organizers to ensure that they do not overstrain themselves.

Physical work may be less physiologically healthy than controlled physical exercises but it provides something exercise cannot – experience of the physical practicalities of life. For the office worker, priest, psychiatrist, architect, teacher and for all those other trades, vocations and professions which do not require the use of the whole body, manual work therefore brings an opportunity to balance their one-sidedness, in a real world context. For the health of the whole human being it can contribute more than just physical fitness.

Building work is more than just physical work. It is work altering the surroundings, making form, making something useful and durable, perhaps even to outlast one's grandchildren. From children's dens to graffiti on walls, eccentric monuments to prestige buildings, there is a widespread desire to change the surroundings, make one's mark on the world, create something. Frustrated, this can distort into destructive forms, but encouraged it has within it the seeds of something artistic.

For the individual, the need to create; for the world, the need for beautiful creative deeds – these are too important to be circumscribed by outdated traditonal roles. Yet in Britain, but not Scandinavia, hardly any women work on building sites. Building is widely thought of as men's work. On the volunteer sites that I have worked on the balance of men and women has been not far from

Working on Nant-y-Cwm Steiner school. Sometimes there were mostly
men on site, sometimes mostly women.

equal. Sometimes the men outnumbered the women, sometimes the women the men.

Every sort of workplace needs this balance. It checks the coarseness characteristic of men-only establishments and brings a balance of outlook which is otherwise easy to lose without one being aware of it. Building materials are typically packaged in volumes too heavy for most women – and indeed for the health of many men – but there is a lot more to building work than just labouring. There is always something for the unskilled or the un-strong to do.

AS WITH other crafts, building can be developed for therapeutic purposes. I have heard of communities in Switzerland working on the rehabilitation of drug abusers which include building as part of their therapeutic programme. Work is arranged in month-long blocks of gardening, cooking and building. Individuals work at each in turn, their time-blocks overlapping, so that each activity is always going on and always has a core of people familiar with the job.

Gardening, cooking and building are all activities in which one's work tangibly creates something. They are the three activities that form the underlying framework of the archetypal home, but are all too often absent from daily life, where buildings are finished objects that one cannot imagine in any other state, food is packaged in such a way as to obscure its agricultural origins, and cooking is reduced to heating.

Yet in all three, both the work process and the final product can be an enjoyable, balancing, grounding and aesthetic experience. They can bring into harmonious relationship individual and community, people and surroundings, body and spirit, life and substance. By this means they can nourish and heal.

In different ways they are working with life. Building encloses and enables living activities and is itself enlivened and continually altered thereby, both in substance and in spirit. Gardening is working with the forces of life in nature, and cooking works further with these for human nutrition.

103

The nucleus of the group who built this retreat-centre had cut their building teeth as volunteers at the Steiner school (Chapter 4) prior to working professionally.

As distinct from cooking and gardening however, building is work for the future, work for others who stand outside the community that you know. In this sense, all building is gift to others, but it can be executed either in the spirit of gift or of obligation. Although the work is given to the future, the effect of that gift in the present is to build and hold together communities. Newcomers to Nant-y-cwm Steiner School frequently comment on the clear link between the amount of work given to support the school and the strength of community that has arisen.

Building can be more than just a job; it can also be enjoyable. It can be educational – it is, for instance, part of every Steiner school curriculum. It can be personally fulfilling, useful and artistic, and it can be developed to become a kind of therapy.

—————IV—————

FRAMEWORKS OF WORKING RELATIONSHIPS

────11────

Responsibilities

IN A gift-work situation there is no buffer of monetary compensation, so relationships, responsibilities and attitudes all have to be dealt with in a more conscious way than elsewhere.

Many initiatives depend upon cheap buildings to get going in the first place. It is often this that influences people to do the work themselves. But if the only concern is just saving money, the potential artistic, social and personal benefits inherent in gift-work will not be realized. Without these, it is hard to sustain projects through protracted demanding periods.

The economic motive may be enough when building for *oneself* – as the revival of interest in self-build homes shows – but buildings for charities and the like can be much bigger. The amount and duration of work and responsibility can be altogether too onerous if it does not, in its turn, bring nourishment. However essential saving money may be, it is not in itself nourishing.

I do not believe that cheapness means ugliness, but if it is elevated from an unavoidable constraint to a dominating priority, it is bound to lead to aesthetic abuses, denying artistic fulfilment to the people working on the project and undervaluing and exploiting them.

In the same way that capitalism, for all its efficiency, reduces the value of a person to his or her economic worth, so does the pursuit of savings alone reduce the free gift of the volunteer to be merely a means of saving money. As in so many other fields, scale is a critical factor. Just as lax kitchen hygiene may be tolerable for an individual, but for a large group risks an epidemic, so problems too small to notice on a small site threaten disaster on a large one.

> When they were working with their own hands, men used to beautify everything they made. Even if it was a warship it was carved with the most fantastic designs because man was interacting with the wood. But machinery does not care for beauty . . .
>
> The cheaper we build, the more beauty we should add to respect man. When man built on his own he used to beautify everything with his own hands. When architects build for the poor, what do we give them from the aesthetic point of view?
>
> Dr Hassan Fathy, 'People's architecture', *People and Planet: Alternative Nobel Prize Speeches* ed Tom Woodhouse, Green Books, 1980.

In gift-work projects ways of doing things that we take for granted in the world around us may just not be tenable. Here, every action depends upon the will to give whereas conventionally, for reasons of material gain, we accept the obligation to do things. This certainly does not apply to all paid work, but wherever personal gain is more important than gift, it does.

Different ways of doing things, inspired by different values, can set in motion processes quite different from the accepted norm. In conventional building, the project at a certain stage becomes the subject of a contract in which the contractor's desire to maximize profit is balanced against the owner's desire to minimize expenditure. The architect gives form to his client's intentions and seeks to ensure that the resulting idea-forms are reproduced in durable substance by the contractor at an agreed price. The contractor seeks to undertake these contractual responsibilities in a manner that will ensure profitability and the foreman manages the workforce for the contractor, with this object in mind. The whole process is one of descent from the realm of inspiration into the realm where monetary considerations are paramount.

With a building constructed by volunteers, the project at this same stage passes into the often unskilled hands of those who will raise it, both in substance and in spirit. Based, as it is, upon giving rather than taking, upon raising matter by working on it artistically – in other words with spiritual values – it is the quite opposite process. Inevitably, it requires wholly transformed relationships between the different groups and individuals involved. After all, giving orders may achieve results in the conventional world, but volunteers receiving them may decide not to give their time any more!

In a volunteer project, the charity, or whatever, – although perhaps having no money at the outset – has a clearly visible need for a building. The architect must find a form appropriate to that need which can be developed at a rate compatible with cash flow and which is able to generate enthusiasm. The builders must work within these limits but also need the opportunity to give and develop their abilities – to participate in a complete way. The foreman's responsibilities are not limited to construction, site efficiency and safety, but extend to listening to the unspoken needs of every individual and trying to find the opportunity for their enrichment and fulfilment.

What, practically, does this require of the different people involved?

THE STEERING GROUP'S RESPONSIBILITIES

IN A charitable initiative, the steering group, or whatever name the people who decide what is to be done, go under, is at the point at which incoming donations – of time and effort as well as money – are transformed into outgoing deeds. I know of no charity in which the amount of donations is equal to the size of the vision.

When you are convinced of the importance of a project but chronically short of money, you naturally want the little money there is to go as far as possible. The trouble is that there is a risk of thinking of volunteer labour as a free gift of time, useful to save money. To regard time, which is part of life, as interchangeable with lifeless money is to value individuals only for their material achievements. However worthy the motive or desperate the need, this attitude is one of *taking* a gift. Taking, not receiving. And this in turn can so easily lead to wanting to take that which is not offered – putting pressure upon those who do not give and taking for granted those who do. Receiving gifts places a heavy burden of responsibility on the recipient – to give in turn at least as much as is received. Giving meals and such like serves as a gesture of appreciation more significant than the actual substance. Most important however is to support the process by which the gift of time and energy which volunteers give contributes to their own personal growth.

All work tends to be tiring, but what makes it nourishing rather than exhausting is the feeling that it is for good purpose, is effective, and is artistically satisfying. Such work involves the whole human being. Situations which don't, imply that the value of people as whole human beings is less than the effects of their one-sided – and therefore energy-draining – work.

Background decisions encourage or obstruct the visibility of good purpose, effectiveness and artistic satisfaction. I have seen precious hours wasted building stairs out of woodwormy timber simply because no new was affordable. Sometimes it is better to spend money on new materials, hired plant or basic tools than to save but waste much time given by many people. On the few occasions when that was possible financially, it has always given progress, and with it the mood on site, a big boost.

Money, time and effort are indissolubly linked. The amount of effort that goes into a job is the result of free decisions; the organizers can't dictate it. Less money therefore is bound to mean longer time. These delays can cause frustration-tensions which can only be relieved by speeding the job up – on occasions even by employing people! Short-term savings therefore need to be seen in a long-term context or sometimes they are not savings at all.

Sometimes it is not a matter of misplaced over-economy, but that the money just isn't there. In this case it is better to switch to jobs with low material costs. Here the price is likely to be in overall productivity, as doing jobs in the wrong order may well make the whole project more chaotic, slower and harder to manage. In the kindergarten project, for instance, the drains, septic tank and underground services were too expensive to do at the beginning. Leaving them till last means saving the interest on some £2,000 of borrowed money, but at a price of having to move sand and gravel heaps, losing material and messing up the work-yard in the process.

There are, in addition, opportunities for the project organizers to provide additional cultural nourishment, of the kind that really transforms a building site and the building workers themselves. I vividly remember one rain-lashed day at Nant-y-Cwm when, in a windowless, ceilingless room, full of demolition rubble and wet cement, we builders sat and listened to a lunchtime music recital by a visiting classical musician. The mood of the moment transformed the rain-spray-soaked room into something worth working for.

Recognition of the need for balance has led to a structure for Nant-y-Cwm building workshops which offers at least as much in the way of artistic, cultural and recreational activities as is received in work – even though it is the practical work which participants usually say has been the most rewarding experience for them.

It is this that is central to the steering-group's responsibility – to give more than it receives and be seen to be doing so. And with volunteer work it receives a lot!

THE ARCHITECT'S RESPONSIBILITY

TO BE architect for a volunteer group is not always easy. Issues, often unseen in other working situations, arise as strong forces. To draw out and weave together their positive aspects needs a delicate path between opposing extremes.

To inspire others you need to present ideas sufficiently concretely formed that they can be experienced as a reality just around the corner, but on the other hand people quite naturally want to change things right through the building stage and into the period of use. And so they should of course, because as initiatives develop, buildings become tangible substance and new individuals become involved, the situation, the needs and the priorities inevitably change. I recently read an architects' report on a college in which they noted that none of the rooms was used for the purpose originally intended – and in this college the activities all had requirements for specific architectural qualities.

It is not easy to retain this flexibility when it threatens favourite areas – to cut up a hall into classrooms for instance. Nor is it easy, when one has gone to such trouble to develop a design in great

113

detail, to find that its real function is to be only a starting point for the imagination of others.

Unlimited flexibility open to the whims of a changing body of users, builders and other concerned people leads nowhere unless it is within the context of an overview. Part of the architect's function is to be an overviewer. For all the frequent need to present realistic sketches as though the design were fixed, overviewing is not the same as master planning.

Within the social forms of a gift-work community, it just does not work to hand down a design and expect others to get on with it. Moreover, the more rigid the plan, the more any fresh suggestions can be experienced as critical attacks. They are no longer gifts to carry a project forward, but negative complaints – all too easily leading in turn to defensive argument.

I prefer a strategy of open-option planning. If a building goes here, how does it set the mood for development in this area? What options does it close? Temporary buildings occupy sites that might subsequently suit permanent buildings, but as everything temporary has a tendency to become indispensable, they are rarely demolished. Piecemeal extensions and adaptations may satisfy the urgent needs of the moment, but over the years the options they have foreclosed can come to be greatly regretted. An evolving overview can help distinguish which options should remain open at any particular stage of development. Flexibility is also needed to be able to listen to the needs of changed situations, and to the new opportunities which only become visible as the building grows.

As the building becomes substance, people see what it means for them. Individual experience however tends to be localized. The carpenter can best see how to construct something; the cook what has to be at hand, where, in the kitchen; the teacher how the light should illuminate the blackboard and children's paper; the caretaker which bits are prone to damage though accident, misuse, play or deliberate vandalism. Unfortunately they rarely see these now glaringly obvious points until the thing in question has either been built or nearly so: all the fixings, services, etc, are in place. It isn't always the right course to change things at a late stage but it is essential to entertain the possibility.

When the detailed design for the Steiner kindergarten was

complete, an opportunity arose to buy additional land and it was suggested to me that the new site would be better. My heart fell at the prospect of starting all over again with the design – for design must respond to site and cannot satisfactorily be transposed. Fortunately the others I consulted shared my (unvoiced) opinion that the advantages of the former site outweighed those of the latter.

There have been other occasions when the fresh eyes of others or my own long involvement in the design of a building have sharpened my attunement to the point when I suddenly recognized something I should have seen earlier – for instance in one project, that the entrance route suggested exclusiveness to the tentative newcomer. Clients may not like the delay incurred by drastic replanning and their confidence in the architect can be shaken if he admits fallibility, but that is poor reason to build upon an unsatisfactory foundation.

As the building grows, it breaks from the limitations of paper. The imagination is no longer circumscribed by the limits of visualization inherent in two-dimensional planning. Now, at last, we can see how a window can speak to a wall, a doorway to a passageway, a sunbeam to a space.

To recognize and develop these conversations requires the flexibility to adapt the design as the building grows, to let the building come alive. It also requires imagination. Imagination has composite roots in flexibility of mind, openness, width of experience and (most essentially) attunement to the situation. For in the situation lies both question and answer.

Attunement varies widely. One hopes, it is most developed in architects. That's what they get chosen for. But there are certainly some who develop or imitate stylistic forms or prefer to make their own personal expressions, rather than serving that quality which is emerging as work progresses and which gives things a wholeness and a spirit. Attunement *is* unequal, none the less it is never right to spurn or deviously manoeuvre others' contributions. Ways need to be found to work together, not as competing pressure groups. When, together, you listen to the being of a project as it grows, it is easy to work together. The problem is that this takes time and for many people other matters take priority.

The greater the enthusiasm that can be generated, the easier it is to upgrade the priority of early, positive, participation.

Even the most conventional recognize that an important aspect of the architect's task is to generate enthusiasm. The thousand-pound model is not to grace the director's waiting room, but to impress the merchant bankers. I have been struck by the change of mood at client meetings when I have brought a model – even a rather poorly made one, as I use models as design tools, continually cutting bits off and sticking new ones on. The building suddenly becomes a reality to the clients. It is, however, a deceptive reality. Even my crude models disguise many elements that show up all too negatively in finished buildings. Still more so do expensive presentation models in their perspex cases. The disguise is never intentional, but the enthusiasm generated by models and other presentations of finished design such as video films, can obscure points which become glaringly obvious in the real situation. It is another reason for openness to adapting the building at full scale.

Enthusiasm is fundamental to gift-work. It carries work forward. But to raise work, enthusiasm must be transformed into inspiration. Enthusiasm depends upon the capacity to accept the gifts of others, so that all can meaningfully participate. Inspiration depends upon recognizing and working with a high aim – and this depends upon attunement. Enthusiasm and inspiration belong together, but their underlying requirements, participation and attunement, can so easily diverge into conflict.

Aspects of building operation and performance are relatively easy to identify and discuss. Aesthetics are not. But if the project as a whole is to be raised to the level of art, to exert a health-giving influence on its users, then aesthetic co-ordination, attunement to a single underlying spirit, however diverse each individual form, is of the utmost importance. So often the single spirit which stands behind perhaps varied aesthetic solutions, is obscured by strong currents of personal opinion and preference. Yet aesthetic matters are too important to fall victim to egocentric whims; they need an attuned objectivity in an area that is all too often dismissed as subjective. This is at the heart of the architect's responsibility; to

116

disentangle an objective assessment of a true art from personal aesthetic preference.*

Art is essentially an activity which streams from the individual. While many may appreciate a work of art, it is one person, or one team acting as one being, who creates it. In its essence, art does not arise from rules – it arises from a disciplined individual attunement – and that is what gives it life.

If art is, in essence, an individual activity, it is also true that it is dominated by individualism. This is more than ever true today, when galleries and magazines are dominated by the cult of individual styles. While it may be clear to many that style may be no more than a set of rules to identify their author it is not so clear where the boundary between individualism and individual attunement lies. Yet there is a boundary, even if we cannot say exactly where. Individualism threatens the whole enterprise of gift-work, with the danger of profiting from the work given by others by indulging personal aspirations. The question always facing the architect must be: Is this my choice because I like it? Because it is my idea? Or, *on the basis of my experience* is this *essentially right?* Is it the most suitable choice in the circumstances?

Sometimes, of course, the architect's objective assessments may be in conflict with the desires of others. If it is a practical matter it is usually possible to provide convincing evidence. 'A window here will cause afternoon glare. If you don't believe me, go and look at such and such an example. Is this price in discomfort acceptable or not?'

It was easy to be convinced of the objective correctness of my stand when I refused to accept a hot-water cylinder placed in front of a window. 'It's only a small one; you won't notice it. Nobody looks out of that window anyway', the plumber said. In such a case I felt that the gift of someone's work could not be accepted in this form.

More recently, during the construction of a chapel, the client liked the appearance of the exposed rafters on the conical ceiling. So

---

* A subject too large for this book. I go into this in detail in *Places of the soul: Architecture and environmental design as a health-giving art* to be published by Thorsons in 1990.

did I, but I felt they were more appropriate to the warmth of a living-room than to the meditative calm of a chapel. The building had brought something to be visible that could not previously have been seen from the drawings, but I felt it necessary to step beyond attractiveness, and to work with the effects of architecture on the spirit. (We agreed in the end.)

When architects take the view that in some matters they alone are right, they also take on great responsibilities, for this attitude comes very close to high-handedness. Inflexibility on matters of principle needs to be more than balanced by humility and flexibility elsewhere – something that is not easy. I rarely manage even to come close to it, and it is made harder by clients, builders and others who interpret flexibility as weakness.

It is my opinion, that in every aspect of life, most conflict is due to differences of opinion and preference. In such cases compromise is the only appropriate action. Only occasionally are matters of true and clear principle involved. In such cases we all know in our inmost being that we cannot live with ourselves if we accept any compromise whatever. Unfortunately I, and I think many other people tend to stand firm on inconsequential issues yet in other matters compromise principles.

Many tasks fall upon the architect of a gift-work project that would not otherwise be asked for. One must explain why things are planned as they are in order to give a larger picture to those who may only work on fragments and for short periods of time. One must cultivate the entunement of those who work on the building, and in the process bring a unified theme to the minds of the contributors. One must find a way to accept contributions of whatever form and maturity without rebuff to the individual or discordance to the whole. Above all one must hold an imagination of the function, the product and the process that can inspire others, for only through inspiration can the material of a building be raised to become art.

It is this very act of raising substance from matter to art that opens the door to acute conflicts of motive within the individual. It is here in his or her inner self, that the architect must face this issue – the distinction between individual activity and individualism. It is

118

the most crucial and potentially destructive issue, yet without it there would be no art. It cannot be sidestepped.

This is the architect's responsibility.

THE FOREMAN'S RESPONSIBILITY

FOREMANSHIP* of a volunteer site is a demanding task. It is not enough just to know every job sufficiently well to be able to demonstrate and teach it. There are also likely to be a lot of uncertainties that don't exist on a conventional site yet which must be accommodated.

I rarely know with certainty how many (or even if any) people will be coming to the next work session. Before a project starts all you have is promises. It is wise in planning anything to divide this by a 'drop-out factor'. (This isn't a fixed percentage; experience of the local community guides one in guessing.) As you get to know the individuals you get to know whose word is never broken, whose always is, who is punctual, who starts at lunchtime, and all the positions in between.

Nor can I know exactly what level of skills or specialisms, if any, there may be on site at any time. I have had a professional carpenter carefully nail a cill plate though roof flashings, in such a way as to create a significant leak, invisible until you notice the results as rot in structural timbers. There have also been people of much less experience who have taught me something almost every time they pick up a tool, and others, completely 'unskilled' (in the trade qualification sense), who can learn unfamiliar tasks fast and work to a high standard. In general, though, some people obviously know more or less what to do, some obviously don't. But I don't know beforehand how many of which sort will come.

It can happen that there are too few people to suit certain jobs, particular specialized skills it would be a pity to waste, or too many unskilled people to supervise on complicated tasks. Sometimes it is better to deflect from urgent tasks so as to find appropriate productive work, sometimes to find undemanding jobs peripheral to the main task.

* I use the word 'foreman' to describe a role equally open to both women and men, without requiring any suppression of female (or male) qualities.

Sometimes there are stocks of perishable materials to use up. Changes in volunteer numbers, in weather, or in priorities may mean that something previously needed in a few days won't be used for months. Some things (like plaster) won't keep for long. Others such as joinery are easily damaged. Advance ordering ensures against delays in delivery but risks deterioration or misuse if materials are on site too far ahead of time. In any case I don't like to wait until the last minute to get something. I want it there when I need it. I therefore keep a list of all small items that will be needed in the near future (or later if there will be problems in obtaining them) and include these each time I telephone order something big enough to be delivered. I have learned the hard way how much time can be spent going to builders' merchants – all the more frustrating if all you want is a dozen screws! I have also learned that sometimes orders must be put off till the beginning of the next month – giving another month to find the money but needing to switch jobs on site when materials run out. At times, regardless of how urgent other tasks are, lack of money may only permit work to be concentrated on high-labour, low-material cost jobs.

I always try to keep a number of options open: jobs sufficiently prepared and supplied for a sudden influx of unskilled labour unfamiliar with the site, indoor jobs for wet weather, outdoor for fine. People who have freely given their time to work are less keen on working inside in beautiful weather, or outside in foul, than those who are paid to work regardless of conditions and are well-equipped with waterproof clothing.

I take the view that it is better to lead by example than by orders, but even here there can be problems. Sometimes things just need to be done, however unpleasant the weather. I try to notice if people feel something is too unsafe, dirty, miserable or unhealthy for them, but I don't always notice in time. Even in answer to questions, some people are unwilling to admit that they have reached their threshold of caution or discomfort.

Thinking ahead means ordering, programming jobs so as not to undo premature work and most especially, working ahead. Many simple jobs that can easily be undertaken by a group of unfamiliar unskilled people need to be preceded by complicated or skilled setting-out or preparatory work. The key to smooth uninterrupted work is to have these preparatory jobs completed ahead of time.

Working ahead can involve just one or two short tasks in one place, like demonstrations of how to fix shaped ceiling framework. It can be the ongoing, leading edge of a job, like slating a valley so that others can do the simpler slating on the rest of the roof. It can range over the whole area of the building like all the complications involved in stepped damp-proof courses along with wall insulation, weep holes, underfloor ventilation and shaping the meeting between wall and ground. To do this all in one would hold everything else up so I try to do a section at a time. Other people aren't held up, but it can be awfully complicated to keep the whole in mind while working on small separate sections!

For much of the blockwork the Steiner kindergarten, because of its complexity, required me to work in advance, only able to use one or two assistants on my own particular tasks, for much of the first quarter of the job. As numbers, skills, weather and delivery of materials all have an element of unpredictability, it is necessary to prepare a fairly wide range of jobs to suit any contingency. A site full of 'prepared jobs' is not the same as one littered with 'unfinished jobs'. The latter leaves the complicated bits unfinished, the former, only the straightforward ones. Nothing is as demoralizing as a site dominated by unfinished jobs – which are very difficult for anyone to successfully pick up and finish off. On the other hand, a site of prepared jobs is a site ready for work to start – ready to receive whatever gift of work and expertise any one may wish to give.

My own foremanship is invariably split between instruction, checking, and working ahead on all those little fiddles that are easier to do than to explain. This of course, is the opposite of the normal trade practice, which is to start one job and complete it before starting another, entailing working indoors in beautiful weather, and sitting unproductively waiting for rain to stop in bad weather. This conventional way of working is easier to supervise and less stressful, but, to volunteers who are keen to exchange their time for something achieved, it is frustrating.

Voluntary and contract sites need to be run in different ways. When, for a short period, sympathetic builders were brought in to speed up volunteer work, their first action was to simplify the job. They completed all the work prepared and kept in readiness for unskilled volunteers. Those who did then come to work found neither job nor materials prepared for them nor anyone with time and interest to instruct them beyond sweeping up the floor, painting the noticeboard, and suchlike odd jobs. When the money ran out and the builders departed, the site was left with all sorts of unfinished fiddles, which no one else wanted to take on. They were not irresponsible at all – on a contract site the normal way to do things is in simple sequence, whereas on a volunteer one it is multi-track preparation.

Foremanship inevitably involves a certain amount of checking. Ideally the foreman is present when work is going on, so that nothing goes wrong, and potential mistakes are spotted in advance. If not the foreman, then every team should have someone who knows the job and can do this. In practice however, there are other matters to attend to, telephone calls to make, materials to measure up, anticipatory jobs to do which demand full attention, and so on. And in the meantime things get done which are incorrect, perhaps even structurally or constructionally unsound.

What am I to do when I find something wrong? It is no light matter to tell someone to take work down. It may only be a small part of the building, but perhaps a whole day's work for someone – work that they are proud of, work which they have given. The first question is: can the mistake be satisfactorily absorbed, for instance by redesign, or duplicating the element? Does it threaten the structural stability, constructional performance or any crucial quality of the building? If it does, can it be put right without demolition? This is a very difficult path to tread. To one side lies unacceptable bodging, to the other, demolition – and with it the rejection – of somebody's freely given gift of time and work. Not only may that person be deeply offended, but also quite possibly will never come to work again! The person who forgot to put cement in the mortar did come back, the one who cut a crucial beam too short didn't!

These are not easy decisions to make as the personal factors can be so sensitive. I certainly have taken down work that could have been left standing, and left standing work that should have been taken down! For the former I am ashamed, for the latter I lose sleep.

Not infrequently foremanship involves teaching the inexperienced without undue cost to the progression of the building. Teaching often moves by subtle degrees from peripheral helping to encouraging someone to do something unsupervised. So, after giving a picture of the project as a whole, I might ask somebody to serve me with mortar, blocks and bricks, progressing to cutting blocks, to placing them on mortar I have laid and checking that they are vertical, to laying their own blocks. Eventually I can leave them on their own for straightforward work. If I know who is coming I can usually choose a job suitable for them. If, on the other hand I am in the middle of something complicated when someone turns up and wants to help, it is harder to find them this sort of work, and I may have to resort to odd jobs. Nobody has actually objected yet, but I feel it much better to give people work where they can see the results, so that rather than 'I mixed cement all day', they can say 'I (or even better, We) built this bit of wall.' It is companionable work, through which they develop new skills, abilities and sensitivities.

To every job there is an appropriate standard of skill, speed of work and quality of material. It is ridiculous to use joinery grade timber for shuttering for concrete, where log-trimmings or woodwormy, damaged, second-hand wood will do. For some jobs wood can be cut with a hammer, for others the sawcuts should be planed.

Speed and efficiency on site require work to be neither below nor above the appropriate level of accuracy, as it should be neither sub-standard nor too slow. Trainees however need to learn how to do things well, *before* they learn to speed up. This is one area of potential conflict between the needs of the building and those of the individuals working on it.

Nowadays, however, I take the view that however slowly somebody does something, it is faster than if they didn't do it – so long as they are not in the way! For such people therefore I try to find jobs which will not hold others up; ones for instance that are not part of a critical sequence, nor astride a movement route, and where without disrupting my own work unduly, I can keep an eye

Not all building work requires skill or strength. If unskilled people come on site and want to help, there is usually something within their abilities.

on them. This gives them the time they need but which I or the building do not have.

The twin keys to good morale on site are individual fulfilment, and the visible acceptably rapid progress of the job. Productivity and fulfilment may well have conflicting requirements. The unskilled are most effectively employed on unskilled jobs, and labouring goes fastest with strong men. But this is clearly not the way to fulfilment, to raise work from drudgery to a healing wholeness.

Ultimately the foreman's responsibility and concern for the building is *through* the people working on it. The quality of the building, visible and tangible, will reflect the concern in this living, social sphere. Beyond work with the material substance of the building, the task is to listen to the unspoken needs of the individuals participating, giving work or introducing new skills or responsibilities at the level each is capable of. The responsibility is both to the building and to the people working, but neither at the expense of the other. This is as much a practical as a moral responsibility, as if you do not care much about the building, inspiration slumps and there soon are no people. And if you don't care about the people, their enthusiasm and standards fall and the building suffers.

In essence the foreman's responsibility is one of balance, finding the appropriate middle way between conflicting requirements and fusing disparate themes together in creative harmony. Which is more important: to train someone in, say, carpentry or to do it faster oneself and use their lack of skill for some unskilled job? The former will invariably slow the speed of work, the latter enhance it. Which does the more damage: delay or boredom? Delay can bring pressures, but boredom loses volunteers! I usually try to keep everybody busy, so carrying the project as a whole forward and allowing everyone to feel that they are meaningfully contributing, but should I concentrate more on urgent priorities, perhaps abandoning other volunteers to cleaning up, tea-making and other unskilled jobs?

As foremanship is the mid-point between building and people, so also it is pivoted between the past and the future, in checking past work and laying the invisible foundations for the smooth running of future work.

Of course, all this is too much to ask of any foreman! Generally problems only arise one at a time. If they can be dealt with one at a time - preferably before they become critical - they can be managed. If many problems simmer together, unnoticed and undealt with, a moment can come when they all set each other off. The foreman does not need to be exceptionally able, but does need to be one jump ahead.

Foremanship of a volunteer group is not easy, but like everything demanding it is either intolerably exhausting or immensely fulfilling. Only by bringing the right personal attitude into an harmonious relationship with the situation can it transform the destructive into the nourishing.

THE RESPONSIBILITIES OF THE VOLUNTEER

TO VOLUNTEER means to give in freedom. Once obligations are even implied, work is no longer truly voluntary. The trouble is that there are lots of good reasons for wishing to impose obligations. I want to know that when somebody has agreed to do something, it will be done, that when somebody has taken one of my tools that it will be returned. In this case people are bound by their own word, given in freedom.

I also want people to clean-up before going home, to be responsible about letting me know whether we need to order anything, rather than just using up the last bag of cement. I want to be sure that when I have explained how to do something - and why - it will be done in that way.

With gift-work orders are out of the question. Work by order means the opposite of freedom, gift and equality. Self-discipline and reliability vary widely amongst individuals, but goodwill, unified by common aim and strengthened by example, is worth a lot. No project could continue without it.

All work given is *gift*. This gives rise to manifold problems, notably an inherent tendency to unreliability as there are none of the penal or reward safeguards that the commercial world depends upon. Yet this is the source of the project's greatest strength, as all,

126

from founders to humblest part-time worker, are working together, whereas in the commercial world, each stratum is not infrequently in conflict with the others. In the building industry open disputes between client, architect, contractors and subcontractors, and between contractors and workforce are by no means rare. That the work is given as gift is its strength – and for that reason, not to mention any moral responsibilities, it must be received as gift, not taken as a due.

Work may be voluntary, but as the initiative progresses, so the organizers become increasingly dependent on the volunteers, especially on a limited number of key figures. This dependency places a responsibility upon them which limits their freedom to withdraw or relax their attendance. All too easily responsibility begins to acquire shades of duty and volunteer work becomes obligation. Yet if personal responsibility exists it dwells naturally within the individual, and is the product, not of necessity but of love. It cannot successfully be imposed.

To progress smoothly, the project may require reliable continuity, but how can this be achieved without compromising the freedom of the volunteers? It seems that only those who hold the ideal strongly will prolong their commitment through varying adversities and life circumstances. They put the project higher on their list of priorities than others do. Every initiative depends upon a core of such people. The stronger the core, the stronger the project. And the stronger the ideal is held, the stronger is the core.

In a world which is not ruled by gain or fear, we must expect to achieve nothing if we lose sight of our ideals. It is not the volunteers' responsibilities which are the real issue, but the responsibilities of all concerned at every level, to keep the ideal clear and inspiring, the star bright.

Which should it be: a volunteer or contract project?

Some people decide on volunteer building because it is the only way they can afford. Others, because they feel it is the right way to go about things.

Some decide to employ contractors, because that is how it is done. Others, because they want to see the job done properly.

In fact, not every volunteer project is cheaper, nor do contractors always do things properly. The choice needs to be looked at in a new light, and so the problems.

Before any project goes down the contract road these questions must be faced:

1. Is the proportion of expenditure on buildings appropriate to the aims of the charity?
2. Is self-sufficiency in *will* necessary to the integrity of the initiative, or is it sufficient to hand over the massive problems to others – the building to contractors and the expense of that to donors?
3. Is it right that given money is used for contractors' profit?

Before any project goes too far down the voluntary road, these questions must be faced:

1. Is the organizing body entering into the true spirit of giving and fulfilling, that is central to voluntary work, or only seeking to benefit by the cheap results?
2. Is the architect able to distinguish between personal preference and universal values so as to be sufficiently flexible where appropriate, yet hold firmness of principle in the overview?
3. Is the foreman able to balance the conflicting material and social requirements and raise these to be inspirational?
4. However worthy the aims, will the ideals of the project inspire volunteers?
5. Are all committed to listening to the living, changing, reality of the situation, or do they prefer to work out of fixed ideas?

There must be a certain level of conscious positive commitment to these issues or all the problems which will invariably occur, will assume much more severe proportions.

Every problem can become either an opportunity or a crisis. The right foundations are not only a moral, but also a practical necessity.

——12——

Who decides what?

IN MOST building projects, decisions which will have a significant effect on people's lives, are decided by an architect. However much fault may also lie with others, such as owners who demand certain building forms and costs, contractors who cut corners, and occupants who ignore day-to-day maintenance, responsibility for many of the social, aesthetic and environmental abuses of recent times lies with architects. Many had good intentions. The environments they planned, however, did not meet the needs of the users. They were disasters.

Recently there has been a vigorous surge of interest in design participation by users. I used to think that either everybody discusses and agrees something, or that somebody alone decides it. Now I realize that most decisions result *both* from somebody *and* from everybody, and that projects go forward as a consequence of many decisions resulting from a variety of individually distinct themes and localized perceptions.

In large projects, if one is not aware of what is happening, the tides of chaos flood in. Decisions are taken by the wrong people, at the wrong time, in the wrong way and for the wrong reasons. Accountants make what in effect are artistic decisions; users want changes when the building is nearly complete, interest groups want influence and may be prepared to horse-trade. It happens in politics, in industry, and not surprisingly it happens in architecture and building.

In small projects these issues can be so merged that their separate demands are not apparent. Although successful decision-making processes in small projects cannot, without modification, be transferred to larger ones, they can be a good basis for a more conscious examination of what is going on.

The design of a small building is relatively easy. The way I normally go about it is to sit down with my clients for a session, rarely less than five hours long. They have a list of requirements – and often a preconceived mental picture. I try hard, though rarely successfully, to have neither. I try to illustrate with sketches, plans, and discussions, the implications, limitations and potential in their ideas. As the session develops, the original preconceptions fall away and we get nearer to what we all would really like to see. By the end of the meeting I can take away a design which is neither mine nor theirs. It is *our* work – so much not mine alone that once a client refused to pay my bill as he felt that he had designed the house!

For convenience only, to speed up the process, I then like to have the authority to develop this through various minor practical changes, and also to open up new areas of choice whenever I can see them – choices which only users can decide. So the design constantly depends upon the conversation between the two parties, even though it is I who help it along the road. Of course, clients may want something that I would find unacceptable, something ugly or exploitative. This is the time to ask them why they chose to appoint me and whether they would not be better served by someone else who shared rather than opposed their values. Very rarely have I had to do this. Twice, I think.

Everything becomes more complicated with larger buildings and larger client groups. Planning is likely to be too complicated to design on the spot. Sometimes, however, one can get the feel of where particular activities would be most appropriate – not just grand entrances and service access, but workshops or pavement stalls here would bring life to this area, here could be a path enriched by activity, or here a place for a quiet heart to the project – a protected garden, surrounded by buildings with quiet uses. Landscape, topography, buildings, trees and other features, paths, routes and activities already established give one a bit of an idea.

They also can blinker the imagination. It's hard sometimes to imagine a continuation of space when you have to see through a wall. Sometimes it can only be imagined on paper, turning your back to the too-strong existing reality.

After walking round the site with the prospective users we can start a planning session, but it is unlikely that a plan will come out of this. What we need first to ascertain are the organizing principles: the qualitative themes, physical relationships and potential of the site to fulfil these, practically and aesthetically. It is premature to draw more than the most tentative atmospheric sketches and so it is harder to get to the heart of these 'qualitative themes' than with a small building.

On occasions I have tried a colour exercise, where we all paint (in watercolour on damp paper) the colour mood that we feel appropriate. It can be a series of paintings to describe a progression through spaces and activities. It isn't easy to avoid literalistic images of buildings or of the colours one sees them painted with. Indeed it isn't easy at all! But think of two libraries – for example a Victorian study and a modern public library. They can be as different in light, colour, and atmosphere as paintings by Rembrandt and Mondrian.

As ideas become more and more substantial, we can walk around the site and imagine activities and the atmosphere they bring. Then imagine the new boundaries we will be building; the approaches, entrances, views, in and out. We can make this more practical with scale sketches on paper, using many semi-transparent overlays to try out different variations.

However enthusiastic people get, there is still a problem with visualization; many people think they understand plans – until they see the plan built! Quick sketches of what things will look like help, as do mock-ups of sizes. We can chalk sizes out on the floor of rooms people are used to working in, and contract spaces by moving furniture around. It isn't a perfect simulation but it's much better than just quoting numbers: who knows what 5.35 metres is? I don't! So I take a spring-rule to such meetings.

This may not be the usual way to design buildings but there is nothing very difficult about it. What is a problem however is the time it takes, which tends to limit the extent to which people are

131

prepared to get involved. Complex projects also need a lot of slow working out, during which quite new ideas present themselves and because of the delays involved in arranging meetings it is all too easy to make decisions personally which should have been left to a larger group. On balance however, the results from such a process are good.

Designs show everything together but in practice things get built one after the other. As time goes by the later phases of a project might have quite different needs from those envisaged at the outset. I try therefore to design things which can be quite satisfactory if only the first phases get built. Design of the later phases should be suggestive only as they could in fact develop in quite different ways.

Which way will a school, a business, a farm, develop? Numerical growth, economic circumstances, specialist directions and many other factors can only be guessed at. Even planning a house for a family has to take account of this evolving future – not necessarily with movable walls (which like adjustable shelving are rarely ever moved!) but with spaces that can be used flexibly. Not amorphous spaces, but spaces of varied character, which ask to be used – non-deterministic starting points. I was recently involved in the design of a farm. Its previous owner had seen its problem as one of inadequate buildings and, advised by the Ministry of Agriculture, had grouped the answers to all these inadequacies into one building. Yet the greatest need was for a heart to the farm – a farm*yard*, off which activities could open, multi-using the space. This building stood exactly where that *space* should have been (and moreover was the ugliest building I have seen for a long time!). Only with difficulty could the options be opened again!

In two Steiner schools I have advised there were problems of growth. So many children were queuing to enter the kindergartens that serious thought was given to starting additional classes. Where would this lead to? Twice as many children coming up through a school would require parallel classes, with consequent pressure upon its social size. Such a course would close options.

If, on the other hand, a new playgroup were started elsewhere, it could, if circumstances were favourable, develop into a kindergarten, in turn either feeding the existing school or itself developing into one. At every stage options for expansion, consolidation, change of direction or staying on the same stepping-stone, would remain open longer.

Design involves planning for the future and often one doesn't know which way a project will develop. In the design of a farm offering courses in all aspects of rural life for schoolchildren, nobody knows yet the extent to which pottery, baking, juicemaking, woodwork and so on will develop. We decided therefore on the main social facilities with possibilities of further organic growth on a street and courtyard basis. Within the constraints of what is already there, and what has some preparations already laid, new people bringing fresh inputs of skill and areas of interest have considerable room to mould the development. Such a stepping-stone policy allows projects to develop in the direction the situation calls for, rather than fulfilling a plan seen as a whole at the outset and thereafter basically inflexible. Such a strategy makes it easier to look forward to what could happen than back to what is no longer appropriate. Criticisms of what has already happened, however valid, only look backwards. To carry projects forward these insights need to be harnessed as contributions.

This sort of process can require a lot of meetings. I for one prefer to (and often do) just get on with the job. A main problem in participatory design is that it takes a lot of time. Generally, people are prepared to spend a lot of time on their homes – furnishing, decorating, arranging, and a lot of time in taking part in the design process. Where it is a place they work in, they are often too busy to give as much time as is needed, and as time goes by different groups – designers, users, financial controllers, builders – can become more isolated from one another than they realize.

It can also happen that something is agreed, then new people get involved and want something completely different. In the design of two churches for the same denomination the two respective priests asked for exactly the opposite qualities. Similarly, in the design of the Steiner kindergarten the spatial and sunlight requirements of successive teachers were markedly different – sun-filled spaces or dreamy half-light.

Written records may cover the architect when requirements

133

change, but they don't help the users. When users and client are different people, the architect is in a position of split responsibilities. When an owner wants his building sprayed against woodworm for a thirty-year guarantee necessary for a grant, I can advise him that there are health risks to the occupants. I can quote statistics of sufferers from timber preservatives* – 40,000 in Germany – but I cannot quantify the risks in the same exact terms as he can the financial benefits.

Owners and users can have entirely different motives such as profitability versus inhabitability. I used naïvely to believe that the client was always right, that the architect's task was to serve. I now see that architects although *answerable* to clients, are *responsible* to users. I still believe the task is to serve, as it is for all of us in whatever role in whatever job, whether employer or employee. Although I no longer believe that even the user is *always* right, I am certain that *on their own* architects can *never* be right.

Traditional societies did not need architects. Cohesion was maintained by the slow speed of evolution of stereotypes within the constraints of available materials and technology. Today it is quite different!

Users know what they want and what they need. That seems obvious enough. My design discussions with individual clients however lead me to think that these wants and needs are constrained by preconceived solutions and limits of imagination. Often I recognize that I am being asked to provide something a client has seen somewhere and liked. It may not be at all appropriate in a different situation! We have never ended up with the rigidly described plan the clients first told me to draw up, but always with something they liked better.

But the architect's task is more than that of discerning clients' real – as distinct from formed – needs and bringing these into conversation with the situational context.

The building will probably outlast the occupancy of any users whom one talks to. In all probability it will be used in new ways

---

* There are less toxic treatments based on Borax (e.g. Timbor for new timber) or pyrethrum (e.g. Permethrin). The most attractive alternative is heat treatment, common on the continent but not in Britain.

and styles of life. Can it, like Georgian terrace houses, have qualities which transcend the use limitations of any particular age? In this sense, the architect's responsibility is *beyond* the user's. It is the responsibility to act as an artist.

Here lies the potential for massive conflicts of interest. It is this view of himself as artist that has led architects to design appalling social disasters.

To develop this path constructively requires, on the one hand, the practice of the artist as the *listener*, and on the other, a recognition that nothing right ever happens *in isolation*. It is not the architect alone who creates beautiful environments. They arise out of the harmonious meeting of what was there before, what is needed, how it is done and how it is eventually used.

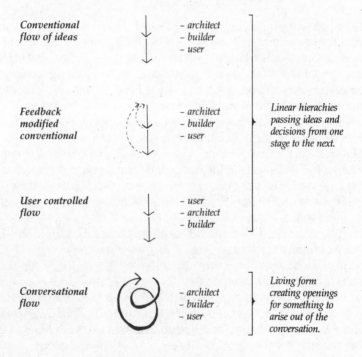

*Conventional
flow of ideas*

 – architect
 – builder
 – user

*Feedback
modified
conventional*

 – architect
 – builder
 – user

*Linear hierachies
passing ideas and
decisions from one
stage to the next.*

**User controlled
flow**

 – user
 – architect
 – builder

*Conversational
flow*

 – architect
 – builder
 – user

*Living form
creating openings
for something to
arise out of the
conversation.*

Flow of ideas.

SOME SEE the architectural stage as the only artistic one, everyone else – builders especially – being philistinic. Yet all the places I have been to that give me a really good feeling, places of spirit, are those where every stage has developed and enhanced the previous one. If we try to develop buildings through *a process of continual improvement* we start to be involved with a wholeness of spirit in place of disparate fragments. Builders and users can be seen as continuing the process of improvement beyond the limitations of paper design. If we consider the whole process by which places and buildings come into being, architects, like planners, builders and users, only take a part in it.

IN A gift-work project people like to feel that they are more than just hands carrying out someone else's fully formed ideas. That is exploitation. Some however feel that giving time and energy buys the right to 'do their own thing'. Once the 'own thing' concept enters, the project is exposed to many divergent pulls. Why should the organizer's, architect's, foreman's ideas be any better than anyone else's? Authoritarian leadership is out of place and counter-productive, but if the organizers can make clear the overall aims and context, the architect the principles underlying both aesthetic and utilitarian aspects of the design, and the foreman the practical reasons for doing things in a particular way, their ideas earn more respect than if they are unduly founded on personal preference and opinion.

Respect or no, there are still people sufficiently convinced of their opinions to want to go off in their own directions. Rigid control is no answer for it not only stifles creativity for the individual, but also fails to take advantage of the many personal contributions – and these are worth a lot. There is here an inherent conflict between art and democracy, for art is traditionally the work and leadership of one, and group work, given as gift, needs democratic forms. Just as users know more about their requirements than do architects, builders know more about how to build things. The more they are involved the less the risk of stupid

136

– and all too easy – oversights and the more satisfactory the results all round.

Medieval cathedrals were built under the inspiring vision of one master builder, the craftsmen enjoying considerable freedom in the design of individual elements. These great buildings were the work of many hands who shared the same vision. This vision was refined by one, usually nameless, master mason who, with his experience and occult knowledge, could raise the building into the realm of sacred mystery and could inspire many to work with him. He had, of course, a name, and it was no doubt well known at the time, but the building is not stamped with it. He gave himself to the building.

To return to a medieval hierarchical pattern would be ridiculous, but the lesson of individuals giving themselves to a shared ideal is eternally relevant. But what is a shared ideal?

If we think of a mountain – how different it looks from different viewpoints, at different seasons, times of day. Yet on every path, how many wonders to discover, unexpected hidden valleys, some stony and bare, some wooded; gentle streams and powerful torrents – all beyond the imagination when we look from one single viewpoint – but all essential elements in the one whole being, one singular reality, the mountain.

Every project has, standing behind it, its own ideal, its own wholeness of mountain. Everyone: designer, craftsman, user, knows only so much of this wholeness, some more, some less than others. The more viewpoints are limited to single, separate perspectives, the less whole, effective and inspiring the work. How can these different viewpoints, including as they sometimes do narrow opinions, sectarian user-interests, or professionally blinkered perceptions, come together to serve a single whole, without being dominated by one individual?

If democracy is not seen as equal opportunity for personal indulgence, but as the recognition of the value of every individual for his own specifically individual contribution, then neither 'doing one's own thing' nor giving orders has any place. Participation can now become a true meeting ground. Such a meeting cannot demand either conformity with the original idea, nor the right to depart freely from it. What is shared is the consistent principle, the ideal, the vision. If this is genuinely shared, the individual variations that arise will all be within the context of a unifying theme. If it isn't they will not, and aesthetic, technical, performance, and ultimately social problems all threaten to arise.

> The healthy social life is found
> when in the mirror of each human soul
> the whole community finds its reflection,
> and when in the community
> the virtue of each one is living.

Rudolf Steiner, 'The motto of the social life', *Verses and Meditations*, Rudolf Steiner Press, London, 1972, p.117.

Conventionally the process of creating buildings is a linear one of delegation of decisions: a simple sequence of commissioning, designing, site management, building and occupying. It is such a well-proven, practical process that we hardly notice that any values of an artistic or inspired nature become reduced somewhere along the line to monetary units. However much money you pour into something, it won't be artistic unless somebody gives him or herself to it. The conventional process works against this.

If on the other hand we can hold an ideal of the building process as one of progressively raising matter into the artistic realm, we have to go about things the opposite way round, so that everyone is artistically involved. First comes the architectural idea. If this is not imposed, but arises out of the needs of users and place – it can have a substance that all can build on. If it is more egocentric in nature, some will ally with it, some oppose it. This requires therefore the involvement of organizers, users and builders. We can work on developing a place in such a way that its needs and the project's needs are symbiotically interwoven. In doing so, ideas form within the group which arise out of the meeting of thought-intentions (such as 'meeting hall') and actual situation.

Individuals need to meet not as proponents of certain ideas, formed separately, but as listeners, already attuned in their minds to the subject. They also need to put aside personal motives, alliances, antipathies and critical judgement, in favour of listening to each idea as it is formed – to what it brings. As Yes – No reactions and advocacy or defence blind us to what can come into being,

138

such a conversation is quite opposite to a debate. From it can arise ideas as living realities, inspired commitment and wise decisions. Even this approach to design as conversation, does not overcome the limitation all design is subject to: that things life-size are different from how they seemed on paper. As the building starts to grow, you can see things you couldn't see before: limitations, improvements, new possibilities. If we don't respond to these, the results will not be as good as they could have been.

> What, specifically, is meant by work here? Certainly not the production of any finished concepts, the amassing of quotes from authoritative sources, the getting-up of a resumé of reading done. Thinking and study engaged in, prior to a meeting, rather serve the purpose of rousing the soul to maximum activity so that it may come into the presence of the spirit, all perception. Work of this sort is a warming up, a brightening of consciousness to render the soul a dwelling place hospitable to insight. One must be willing to sacrifice previous thinking, as one does in the second stage of meditation, in order to clear the scene for fresh illumination.
>
> The principle here is the same as that advanced by Rudolf Steiner when he advised teachers to prepare their lessons painstakingly and then be ready to sacrifice the prepared plan at the dictate of circumstances which may point to an entirely fresh approach to their material. If one is well prepared, he said, one will find the inspiration needed. Indeed, the principle is common to all esoteric striving: Invite the spirit by becoming spiritually active, and then hold yourself open to its visitation.
>
> (Marjorie Spock, *The Art of Goethean Conversation,* Spring Valley, New York, St George Publications, 1983.)

As the building advances, ideas – forms, spaces, colour, light, use, fittings, and so on – which could not be conceived on the flat paper on the drawing board, or even in a model – start to arise. On paper, everything is reduced to two dimensions. The third can only be inferred. Models show form as experienced from the outside. Yet we live in space, a space bounded, defined, articulated by the boundaries of forms, often forms only partially experienced. We live in other words in an environment that the drawings and model do *not* show us. But the unfinished building does, for now we can

139

How could this old stone wall be finished off and the new rise from it? Only by working with the hands do the right stones make the right ledges in the right places. Such a thing can only be designed as a suggestion – it is the action of doing it that carries the design forward. It is the builder who must be the artist.

see the space and its bounding surface. Now we can see, even mock up, how it would be if we did this or that – set a window just here for instance in relation to view, to what one sees on entry to the room, to sunlight, to the frame of the wall and ceiling – and so on, even to what shape the door should have, which way it should open. Only in the half-finished building do the spatial and environmental potentials invisible in the model and drawings become apparent! There are in fact a lot of decisions which are best made on site – although it is wise to have a design on paper first even though you intend to discard it!

Already, at this stage, people can begin to see in some ways what it will be like, how they could use it, arrange furniture and so on. Of course scale and lighting can be very misleading. Scale in particular is deceptive. Buildings look tiny at the excavation stage, large when the walls rise above eye level, small again in the gloom when the roof is on, and spacious if painted white internally prior to any furnishing.

This second stage of user – and builder – design starts as soon as building upwards starts with the layout of the first bricks, which will define the actual shape of rooms. The exact curves of the walls in the Steiner kindergarten for instance, I could only work towards on paper. Curves can be weak and lifeless or firm and alive. Only in placing these bricks could the shapes of these walls come to life.

The process continues as the building grows. The building starts to become its own full-size design model, to develop its own life. The growing building as it were, asks for developments in one direction or another. The more you work on something the more attunement to this situation replaces the desire just to implant personal fresh ideas. Working in this way has profound implications for the people involved in the process, as well as for the building.

I HAVE come to realize that gift-work is sustained by the inspired will. If this inspiration isn't nourished soon there won't be any volunteers! While gift-work can be seen as a one-way process of giving, it actually requires the work situation to give to the

The tree-trunk post and curved beam arose out of discussing with the builder *how* we could achieve the headroom and resolve the structure. The best touches in this building – as in any other – found their form, and their feel, through such conversations.

volunteer. I try therefore to involve people in the reasons behind things and encourage their thoughts and advice as to how to make them. More importantly, I try to cultivate aesthetic involvement, the seeds of which lie within each of us. Building essentially is an artistic process, but on a larger scale, with more constraints and parallel functions than painting a picture or modelling a sculpture. It is out of our aesthetic attunement, that *how* to do something – plaster a wall for instance – emerges. A plaster surface can be dead, limp, without strength, like a lump of jelly – or living, harmonious and firm, like a rock on the seashore. I can describe how to do it in words, but only the experience of actually *doing* it can make it live inside oneself, can enliven the aesthetic sense.

If we work in this way we become aware that we are listening to something that is developing all the time. The building is, as it were, a thought-form incarnating into matter. But it is a thought-form which is alive, is born of the meeting of idealized needs and real situation. It is not something it is possible to rigidly circumscribe at an early stage by a fixed design. This leaves no room for life.

What started as the desire to have a say has become transformed into listening – and giving an answer. Whereas conventional decision-making is often a matter of power, it has in this circumstance become a means of entering into a living process, and by so doing creating life.

Ultimately the right decisions are not made by one person or even by one group. They arise out of the situation. The more closely we can listen to the situation, the better will be the result.

Something new, intangible, but very real, has been born – that which, beyond the individual contributors, comes to life in conversation. The successes I enjoy revisiting are due to this and the failures mark lapses into deafness – egocentric preference rather than listening and responding.

What this means for the building workers is that more even than completing the physical building they are entering into the process of giving birth to a 'spirit of the project'.

Participation, listening and gift, in place of power, imposition and taking can help such a spirit to appear in any kind of group activity.

The crude structural enclosure of walls and rafters ask to be set into song. A framework to bring the planes into conversation, itself metamorphosing from one situation to the next, is the first step. Hand-finished plaster carries the process further.

144

In a building or dramatic performance it is perhaps more visible, but it can exist anywhere.

When I revisit buildings I have designed I notice many details that I would like to have seen done differently, yet without this way of working these buildings would never have got as far as they did. Perhaps now I could design or make them better – but prior to the involvement of others *I never could have.*

—— V ——

ADAPTING PROJECTS TO SUIT GIFT-WORK

─────13─────

Design implications

SOMETIMES PEOPLE say to me: 'We would like to build it ourselves, but we can't.' Building often seems too daunting for someone who doesn't know how to do it, but when we go over their project it's not so hard to identify which bits they could do, and which exceed their confidence. Much of the atmosphere of a building depends upon fittings and finishes and many of these do not require professional skills. Undulating plaster, handmade shelving, second-hand timber counters and cupboards, lazure-painting, can transform an otherwise dull building. All of these can be more sensitively and better done by a committed amateur than by normal skilled building tradesmen. Indeed, I often advise clients who will be employing builders to do certain jobs themselves, because it is they who can do them best. For these jobs, sensitivity is more important than skill.

I also meet people who say: 'We would like it to be built by volunteers.' By volunteers, they mean people other than themselves, people who work for nothing. Sometimes they refer to ET workers. People who will not cost them anything. I try to find out what they mean. Do they mean 'I want something for nothing', or do they mean 'I would like to see this project founded on gift'? Usually it is both. Where, however, the former intention predominates, there are bound to be problems.

Visibly wasted money or grandiose schemes which do not seek to constrain costs by utilizing available resources can give rise to the feeling that volunteer labour is used merely as a cheap way of fulfilling expensive ambitions.

The value of a product – even monetary value – is the result of the combination of materials and craftsmanship. The most expensive materials combined with dull and uncommitted craftsmanship give a feeling of hollowness, emptiness, gaudy ornament, whereas artistic craftsmanship with humble but durable materials, warms the soul. My first choice therefore is to recommend traditional materials and conventional methods of construction for aesthetic reasons. There are other reasons also.

The cost of self-built and volunteer building projects varies widely, ranging from a fraction of contract prices for similar work, to well above it. Design must answer for a large part of this.

Self-builders and volunteers are often thought of as capable only of a low standard of craftsmanship. Although by no means always true, it is prudent to avoid construction which depends upon skills or precision. Some hold that they are fit only to assemble pre-manufactured components. Effectively, the work of building is largely shifted from site to factory and the cost of the components will show this. If there are further problems of timetabling deliveries and damage to goods stored on site, the actual savings can become very small. In any case the contribution of the volunteer is greatly devalued as is the potential for evolutionary design inherent in this way of working.

If, on the other hand, material costs are kept under close control but labour time is maximized, gift-work can produce buildings of a much, much higher standard than contract work, as a considerable amount of time-consuming handwork can be incorporated. Most building work can be carried out adequately with but moderate skill. Design can minimize dependency upon skill or specialized (and expensive) equipment. Structural and constructionally vital elements can tolerate more mistakes if they are oversized.

It is as well, for instance, if walls are stronger by being wider or having more curves, angles or buttressing walls than required. One inch out of vertical is quite a lot for a four-inch wall, not so much for a nine-inch one. Similarly I use nine-inch vertical damp-proof course instead of four-inch where openings close blockwork cavities, to protect against mortar droppings. In practice the extra

150

Why buy a door-latch when you can make one? Why pass unconsciously from one room to another when the action, even of opening and closing the door, can be a delight? A retired shopkeeper and I discussed door-latches and this is what he made.

151

material cost is small. Concrete work may be 90 per cent labouring, but the cost comes in the shuttering. In any case I try to avoid imposing such miserable work on anybody. Where textures of finished surfaces admit to being handmade, the meticulous, experienced skill required for perfect uniform surfaces is not necessary.

The contrast between vernacular and high-technology building exemplifies this. On the one hand, the structural principles are simple and visible. The critical elements are over-sized: walls, for instance, thick enough to be stable, even if out of vertical. This can give a sense of durability and timeless response so welcome in today's environment. The finishes may well be imprecise. It is this which gives texture, richness and delight, in this kind of building. As the building ages, its surface textures are further enriched by weathering, damage and repair. Many such buildings were constructed by people who were not professional builders.

With high-technology building, accurately dimensioned, hard-edged components are skilfully and precisely fitted together, with minimal room for error. Structural design is likewise precise and economical and depends upon accuracy in construction. The finish is smooth and immaculate and is ruined by poor construction, weathering, damage or repair. The damage and deterioration that inevitably accompany ageing are either expensive to correct or leave the building looking squalid and cheap like an old table with a chipped laminate top.

Simplicity in constructional method and robustness of surface appearance have an influence on the design as a whole but they do not define it. Just because the building is traditionally and economically constructed, it does not mean that it has to look conventional or cheap. It can do, but if you go to all the trouble to make something, it seems worth putting a small amount of additional energy and commitment in to make something that it isn't possible to buy. Indeed, to make something better than you can buy.

I sometimes look through manufacturer's catalogues and realize that I couldn't make something at the price at which they sell it. The discount they get on their materials covers their production, distribution, and advertising costs, even allowing for profit and retailers' mark-up.

If you want to build a cheap quick building, all the bits are available, mass-produced. All you have to do is put them together. You save paying for the contractor's time but you do pay more for the materials because he can get better discounts, VAT refunds and take surplus materials on to his next job. What you get at the end will almost certainly be cheaper than you could have bought it for, but not as cheap as you might have hoped for. What a sweat for only a moderate saving and probably an inferior product! I am not prepared to give up a slice of my life just to save money. It's too much work. To my mind, if one chooses the self-build approach it's only worth while if you can take advantage of what it offers. It offers the chance to do everything artistically.

At every stage there is choice, denied by the catalogues. Why should this window be square? What kind of arch form – perhaps asymmetrical to respond to the wall shape in which it is set? Why should the door latch be steel, invisible in the door, and devoid of experience when I use it? Of course, opening up these options also opens up the possibility of unfounded whimsicality. It is necessary also to ask, why *shouldn't* the room, window, balcony be rectangular? Why shouldn't it be a standard manufactured item? Why shouldn't particular items look like they actually do?

When people ask me what sort of design compromises have to be made for unskilled builders, I generally answer 'None'. For self-building is an opportunity more than a limitation.

Gift-work relieves, though it does not eliminate, pressures of costs. It exacerbates, however, pressures of time. The energy and money available tend to be limited with the result that gift-work is likely to be slow. There can be great pressures to compromise aspects of the design to speed things up. Speeded-up building by cutting time-corners tends to lead to maintenance problems. Compromises in design may or may not be noticeable because there is no reference point to compare them to, but each is a sadly wasted opportunity.

Fundamental to the issue of speed is the price of time. If time is reckoned in monetary or acquisitive terms, it denies its real value – for time is gift. Time given is a slice of our lives. It is human spirit contributed. Instant results lack this spiritual content, they depend instead upon technology. It is not beyond imagination to picture computer-aided design relayed directly to a robotic component factory – design and erection taking a matter of days. But what artistic quality, what work therapy and what independence can we expect?

There is another aspect that has a bearing on design, namely the ability of a project to enjoy support. Gift-work depends upon gift-money. Not just because it is hard to get commercial money but because profit-making loans are a noose with strings which those with other motives are able to pull. I have seen banks lend freely to farmers, only to switch policy a few years later. Bankruptcy destroys a whole way of life for a farmer whose forebears have been on the land back into the mists of history. For a charitable initiative bankruptcy is the end. It only has a few material possessions for the liquidator to sell, but they were home to the spirit of the initiative.

A project may borrow plenty of commercial money, but what it depends upon is a backbone of donations and low-interest loans. These are given not because of the far-sightedness of the vision, but because the practical evidence shows that this vision can become a reality. The more work to be seen, the more inspiring and convincing is the project in the eyes of the donors.

Performance characteristics of a building may be constrained by skill and cost, acoustic screens, for instance, are expensive to buy and difficult to make. Artistic variation is only limited by potential difficulties in supervision. The more varied parts of a building are, the less easy it is for the foreman to construct a little bit as an example of what to do somewhere else. Complex buildings tend therefore to need more 'working ahead' than do simple ones, yet it is possible to build forms and spaces of a kind that cannot be adequately worked out on paper.

Subtle forms which cannot be described with lines are virtually impossible to visualize and develop in anything less than the building itself. Similarly, texture, which brings life (or death) to very simple forms defies adequate description in drawing, model or words. Design frozen by contract must accept the risk that things will work out badly – that the complex forms look contrived, that what should be graceful simplicity is sterile and boring. For a really small room, texture can make all the difference between the monk's or the prisoner's cell. Evolutionary design gives better scope to work with these things than does any standardized construction.

Kit assembly leaves little room for improvisation. Modern cars, when they break down, depend upon replacement of their sealed unit components, whereas older ones could be fixed by an ingenious mechanic. Ingenuity, spurred by necessity and developed by constant exercise, can resolve the apparently impossible.

There is plenty of ingenuity practised by self-taught builders. Often, lacking understanding or concern, it takes the form of short-lived, unsightly bodging. But other examples are beautiful, resourceful and functionally effective. One self-build client of mine, to save on the expense of hiring adjustable props used tree-trunks to support a floor when he removed part of a wall. He so liked the effect of a forest in his living-room that he chose timber cut from his hedgerows for every structural post. Handrails, doorhandles and furniture followed, cut from branches. This sort of independence from manufacturers and suppliers can be developed when every detail is unique.

The more out-of-the-ordinary building materials are, the harder they are to find, the fewer are the suppliers and the more dependent the building and its construction programme is upon outside influence. Unfortunately it is industrial products, polluting in manufacture and with adverse long-term effects on the user's health, that are the easiest to obtain. The alternatives – where they exist – can be very hard, if not impossible, to get hold of.

Most insulation blocks for example are made of fuel ash, pumice or foundry slag, emitting three to ten times as much radon as ordinary concrete blocks. For one job I managed, however, to trace a low radioactivity insulation block, but could only buy them in lorry and trailer loads, nearly four times as many as I needed! It is not the bricklayer but the future occupants who will be at risk here. The same problem exists for almost every harmful material. Biologically safe wood preservatives for instance are sold on the continent but until very recently I could not find them in Britain.

In general I try to use healthy materials, especially in areas that will have long periods of occupancy. Rooms used only intermittently such as toilets and storerooms are less demanding. Apart from this I try to use easily available materials and favour those that give the greater independence from suppliers or specialist subcontractors. I try to avoid anything requiring the hire of specialist skills or equipment over a long period. These must not only be paid for but timetabled. Either the subcontractors wait for the building to be ready for them (which is expensive) or the building is held up until they can come (which is frustrating). I try also to minimize 'special materials' – those that need to be specially ordered, carefully stored and protected from misuse until they eventually are built in.

For the building, this all means simple but good materials; traditional, low-skill construction, but freedom for the building form to respond to the situation.

Gift-building does not limit design; it frees it.

The atmosphere of the site

WHEN I think of the average building site, I think of ground churned up by big machinery, constant noise, tough men, coarse language, ugliness and mud. It is an atmosphere of assault. Men and machines attack the landscape to impose buildings upon it.

The unrelieved ugliness of the site is of no consequence economically. Only the end result matters, and the workforce will continue to work and be paid for it, up to that point. But nobody is paying a volunteer workforce. If the site is too ugly, too loud, too coarse, abusive or aggressive, their will to come to work can suffer a severe setback.

A bad atmosphere on site is the foundation of bad morale and declining volunteer members. A good atmosphere is not only essential for the invisible spiritual foundations on which the building is to be constructed but is also a practical necessity.

The mess and mud of building may be largely unavoidable, but seldom totally so. I have been on a number of small contract sites that were much tidier than volunteer sites. In addition to accumulations of oddments (which the thrift born of financial stringency does not let one throw away) there can be the feeling that 'I come here to work', and tidying up is not work. So people work until the last minute, then go. Tidying up is always demoralizing as a way to start the day. The price of maximizing the day's production in this way is squalor and deteriorating morale.

If you can't afford skip hire, where can you put rubbish? Old cement sacks are only occasionally needed to protect brickwork against frost, or fresh concrete against the sun. Some things can find new uses; broken blocks become hardcore, the smaller timber offcuts become kindling for someone's fire at home or a cooking fire on site (but not if they have been preservative treated – there is a risk of arsenic ash or dioxin smoke). Old nails and bits of steel can go into the heat-storage parts of a building, broken glass into soakaways. Plastic drums and containers can be cut into buckets, pipe offcuts saved as they usually come in handy; but what about other bits of plastic? Much of the mess on most sites is disorder, mud and plastic. Disorder can be dealt with by giving things orderly attention. Mud can be used – for raising ground levels or building hedge-banks for instance. But plastic – if you burn it, many sorts make poisonous smoke; if you bury it, it doesn't rot; if you leave it, it may blow away. On a contract project recently, expanded polystyrene sheets incautiously stored were torn to pieces by the wind which spread indestructible bits over the landscape. Autopsy on a neighbour's dead cow found polystyrene in the rumen as the possible cause of death. In general I prefer to avoid plastics.

Looking after a site can be like looking after a home. It is both a cradle of the future and already a place, a moment in its own right.

You can see the effect of care or lack of it as clearly on a half-built site as in a finished building. Maintenance, with none of the creative appeal of building, can be merely a chore, yet if anyone has seen tools, a site, a building, a home, maintained with real love, it is clear that this inner attitude is visible for all to see.

In Scandinavia there is the tradition of the house or farm gnome (*tomten, nissen*). Respected and cared for, he will look after the farm. Abused, he will see that things go wrong. I have never seen one, but I have experienced the deterioration and accidents which defy rational prediction, that befall an unloved farm. I have also seen how they can be reversed once the place is again loved, finds again its spirit.

A plant, an animal, a child, fed and looked after according to the rule book, will survive. But to bloom, it must be cared for with love. The same applies to a building, mineral and lifeless as it appears.

Anxious not to abuse the spirit of the woodland setting, we placed materials within or right beside the building. This, however, together with the complications of curvilinear forms and a steep slope, made it hard to see a coherent whole. We tried therefore to bring the whole building up a level at a time so that it could more easily be experienced. (Steiner kindergarten)

Machinery makes a lot of difference to the atmosphere. On the one hand, noises of machines indicate work proceeding vigorously. As smoke from factory chimneys signals prosperity, so machine noises signal job progress. Some machines such as cement-mixers have big advantages and do little damage, but others however, are just too noisy to be companionable – no one can work near them without fatigue. Some are dangerous to be near. Big machines such as excavators, and heavy lorries can make an incredible mess, turning earth to mud, order to sludge.

Machines make work lighter and the job proceeds much faster. But machinery, by its nature, assaults. Much of this assault can be kept within acceptable levels and is redeemed by the creativity of the work. Some cannot. Progress is important – for the project as a whole and for morale on site – but it is not *all-important*. Progress by assault carries a price, a price in atmosphere on the site, a price to the environment finally created. Usually the price of mechanical support is low, its benefit high. But not always.

With or without machinery, how the site is managed affects the speed of progress. Ideally goods move to their point of use without double handling. I have learnt this the hard way, time and again having to relocate piles of stores whenever they were in the way of access, scaffolding, measurements, taking levels or other work. On a productive site everyone has something to do – a lesson I learnt by either waiting for someone to finish something or get out of the way, of having them waiting for me to describe what to do next. Such a system is dependent on the weakest links – the slowest people.

Teamwork, on the other hand, brings people together in a way that can absorb the less able. It develops relationships where it is natural to develop an awareness of each other's needs. Productivity, and with it morale, inevitably benefit.

Underlying teamwork is the feeling that all are working together for something more worthwhile than individual gain. This sort of unity of inner commitment can help greatly with all the less attractive work which is part of building. Not every building job is interesting, engaging, creative or pleasant. Some are dull, burdensome, exhausting, unhealthy. Health risks can always be reduced, though not always eliminated, by proper precautions, but some jobs will always remain unpleasant. Such jobs have to be done. Indeed, of the whole job one can say 'It has to be done' – and herein lies a crucial issue. On the one hand, this need to have work executed can begin to become an obligation laid upon volunteers. On the other, they give their work as a gift. If the gift is asked for before it is offered, if the demand for the material results tramples the spiritual contribution, the volunteers can feel themselves taken for granted, undervalued – but even worse, because the necessity of the work places chains of obligation upon them, *unfree*.

This knife edge between obligation and freedom, material needs and spiritual gifts, haunts every site. Once seen, the ghost is always safer . . . . But it is not always so easy to see.

—15—

Restructuring tasks to suit mixed-ability teams

HOW CAN the traditional work structure of the building site be modified to best take advantage of and reward given labour with its varying skills and experience, often at a low level?

In normal practice, skilled work is given over to the craftsmen and to the unskilled go all the mindless, menial, burdensome jobs. Unskilled workers are seen as merely the servants of the skilled. Some people enjoy hard physical labour without any demands on their thought, but for others it can be rather demoralizing. For volunteers, it does not feel fair, as all are equal in their gifts of time yet some get interesting, while others, soul-destroying jobs. Nor is it practical as the ratio of unskilled to skilled is invariably out of balance.

For a team, however, the ratio is not critical. We can break down and recombine the components of any tasks to suit the team.

Conventionally, to lay blocks, a skilled bricklayer does all the work on the wall and a labourer supplies him with everything he needs. A common ratio is one labourer to serve two bricklayers.

Quite a large team can lay blocks. Of these, only the one laying out the mortar needs a moderate level of skill. Having only one task, this person can use a trowel in each hand to compensate for lack of skill. Someone else, with reasonable strength and a gentle touch, places the blocks, while another checks them with a spirit level and smoothes in any protruding soft mortar. Others cut blocks, place cavity insulation and set out wall ties. More mortar-

Teamwork makes heavy work lighter, faster and more joyful. Imagine carrying seventeen tons of earth on to a roof on your own?

layers, block-setters and levellers can follow closely with the next course. Another two mix mortar and deliver all the necessary supplies. Most of these jobs can be rotated around the group. If work can proceed like this, movement flows through the team like waves through the sea; the team is moving and working as one being. Progress is markedly rapid and morale is excellent.

Any time-and-motion study, even random photographs, shows how much of the time people are standing around. If standing-around time can be cut down from perhaps say 50 per cent to 5 per cent, then work really shoots ahead – and with it morale.

I have worked (as a carpenter) on a site inadequately supplied with scaffold and ladders. Standing on floor joists to work meant that I could not have my tool bag within reach, but had to place it on the floor below. I spent almost as much time stepping carefully from beam to beam along the building, down the ladder to get a tool and up again, as actually working. And sometimes somebody had moved the ladder! Every change of balance required me to stop and look where to put my foot. Everyone was in the same position, carrying stones one at a time up ladders, hunting out the electric extension lead from where it had last been used, putting up rafters to mark them, but having to take them down and round about the building to find a place to cut them. Everyone on site worked hard, but progress was slow. We worked as individuals, or at best as pairs (sometimes with a third person unproductively on the edge of the group). We did not work as one being and the development and morale of the job showed it.

A conventional working group is limited to the speed of its slowest members. In a team, however, slow or incompetent individuals or those inexperienced enough not to know what is coming next can be carried by the others. The team is limited not by its weakest member, but by weakness in its cohesiveness.

It takes some skill and not a little preparation to set a team into smooth motion, unobstructed by hold-ups and sufficiently flexibly structured to absorb different individual limitations. Simple jobs that would otherwise be boring lend themselves to this invigorating method. Complicated, slow jobs on the other hand are better suited to single craftsmen or craftsman-and-mate pairs. Using the team principle as the starting point to think about work, a job can be broken down into tasks suitable for teams, individuals and pairs.

For teamwork to flow easily, I try to arrange that skilled individuals work ahead to prepare tasks to the stage when a team can manage them effectively. The key to smooth progress is the ability to identify and deal with all the little fiddles well before they become hold-ups. The knack lies in matching the working speeds of individuals and teams in such a way that the right people are free at the right time to deal with all the 'working ahead'.

Individuals range widely in experience and speed of learning. After breaking the job down into team tasks I try to match appropriate teams, and match the teams and individuals with the tasks. Some work is appropriate for unsupervised novices, some requires that the team have a skilled member, some must be finished before the next step can be started.

Volunteer numbers expand or contract but in many cases the team sizes are only flexible within fixed limits. Only so many people can work on demolition without risk of injury from falling material. Excavation requires sufficient space between people for the safe swinging of a pick. There cannot safely be too many people on a roof. No more than two people can fit two pipes together and the preparations – cutting, chamfering, marking, lubricating – only employ one other. For some jobs the team sizes need to vary as the job progresses. Plaster-boarding, for example, tends to start with a maximum of two teams of four per plane of wall or ceiling. More would get in each other's way. After a while smaller teams of two or three can splinter off to deal with all the little fiddly bits that now are needed, and are too small for larger groups.

For most jobs the minimum effective number is two. Two to move awkward, long or heavy objects, one to work and one to carry materials; one above on scaffold or roof and one to pass up supplies. Two is a balancing number. Wherever the task requires balancing functions there must be two people. For ceiling plasterboard a minimum of three and a maximum of four are needed to lift and fix each board without breaking it. Specialized tools or ingenuity with brooms or props can reduce this number but not everyone can contrive such a system. Manual excavation needs a lot of people if it is not to be demoralizingly slow drudgery.

A general picture of the maximum and minimum numbers who can work together, the sequence of work, the skill and supervision requirements, makes it easier to arrange work effectively.

Amongst the advantages of this way of working is the fact that the task can be chosen and often adapted to suit the team with its individual blend of speeds, skills and experience. Completely unskilled people can be involved at a pace that suits them, the rest of the team covering their inexperience. There is also the advantage that the visible accomplishment is often in reverse ratio to the skill required. It is the block-placer who makes the wall rise, while it is the mortar-layer of whom skill is required. Similarly the person who nails timbers into place needs less skill than the person who cuts them, and that person in turn less than the one who marks them. On the whole those who come with lack of confidence in their abilities go home with the feeling of something good achieved. For the more skilled, it is satisfying to set things up for others to experience the accomplishment. The key to successful teamwork is attention to each other. If you know what the next person needs before he or she is held up for it, work proceeds smoothly and fast. If not, everybody ends up hanging around. Not only is this better for work, for morale, and for the whole project but it is also a good and fulfilling social exercise.

Listening to the unspoken and thinking for each other must be at the heart of this kind of work if a relationship of leader and followers is to be avoided. Work by community is really the only appropriate form for volunteer work – and you can feel the difference from hierarchically ordered work in the results! Listening is the only appropriate way to achieve it.

16

Working with second-hand materials

SECOND-HAND materials are often only a quarter of the price of new. As half of the cost of a conventional building is the cost of materials, and voluntary labour is free, very considerable savings can be made. I would not advise buying *everything* second-hand. During a brick strike I bought second-hand bricks from a dynamited factory chimney. But they were all the wrong size to use with the bricks and blocks I already had! On the whole, however, the savings are great and, with labour-intensive buildings, the total cost can be but a fraction of estimated contract cost.

Some consider second-hand materials inferior but acceptable because of their price. Others see in them fortuitous opportunities and adapt their project to accommodate whatever comes to hand such as pub windows or Victorian iron staircases. In the same way that the contents of yesterday's junk shop – or village pond – are today's antiques, wonderful items can be rescued and given a new life, if you have the time and inclination to let the design be moulded by this factor.

There are also the wider environmental considerations. The tree once felled has another kind of life as timber in a building. The longer this is, the longer other trees can live. If you've ever seen felled woodland, quarries or brickfields, you can see how much environmental damage materials cause.

I generally combine new and second-hand materials, favouring one when its special character is appropriate. New timber for joinery, for instance, as I know it is nail-free. I learned this by making windows with excellent quality second-hand wood – but concealed nails are cruel to chisels, planes and best saws. Old floor tiles, irregular in shape and colour, are more attractive than the mechanically uniform modern ones. Second-hand material is usually more awkward to obtain and to work with but cheaper and better for the environment. There are, however, some materials I would *never* re-use – such as demolition rubble for hardcore under buildings, because it may well contain timber, possibly already infected with dry rot; non-armoured electric cable, because it may well have been damaged.

Second-hand materials are not always inferior in quality, but take more time and trouble to get, prepare and work with. And they need to be over-ordered to allow for breakage and damage. Many slates have hairline cracks that do not show up in the pre-purchase inspection. I therefore usually order 20 per cent over the calculated quantity and slate the larger roof areas first. If there aren't enough to finish the job, only the smaller roof areas remain. If the same size aren't obtainable, it is no great loss to re-batten a small roof for a different gauge of slating.

Timber is often notched and has end damage – especially splits. It is therefore often necessary to oversize structural timbers by one inch and order extra boards – inspection of a sample batch gives a clue as to how many. It is surprising how many widths of floorboard come in the same batch, presumably originally from the same building. When a floor is longer than the boards, so that they have to butt end to end, the time taken and the boards discarded to ensure consistent width, is considerable. Edge damage and woodworm, which for some reason seem to find floorboards their favourite food, mean that boards must often be cut short. For these reasons also it is necessary to order more than one expects to use. The greatest savings without the penalty of too much additional work are usually found in the larger structural timbers or those of common general purpose sizes.

Second-hand timber is frequently well studded with nails. Both ends may be damaged. I rarely de-nail, but merely knock over protruding points for safety. Initially I do not clench these down as I may need to extract *one* nail if it is on the line of a saw cut. If it is necessary to cut or joint at both ends, I mark out both before sawing, to attempt to avoid having to pull out nails. Sometimes unavoidable nails won't pull-out or the heads pull off. If they can't be punched through, I saw all around and cut through the wood where the nail is with a hacksaw.

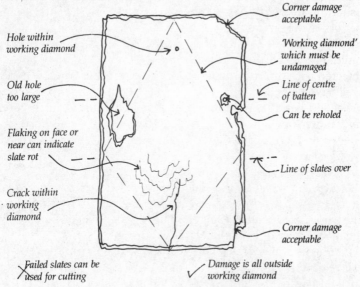

Corner damage acceptable

Hole within working diamond

'Working diamond' which must be undamaged

Old hole too large

Line of centre of batten

Can be reholed

Flaking on face or near can indicate slate rot

Line of slates over

Crack within working diamond

Corner damage acceptable

Failed slates can be used for cutting

Damage is all outside working diamond

Assessing second-hand slates.

As for slates, after visual inspection I hold each loosely and tap lightly with a hammer. If they ring with a clear tone, they pass. If they sound dead they are immediately suspect. A buzz means either a serious crack or a loose flake, which may not matter. Re-inspection usually shows up the fault. If in doubt, wet it; as it dries cracks show up. Discarded slates then go into stacks for longways or crossways cutting. On the roof it is as well to check each slate again before nailing it. With the hammer head I scratch a large cross and circle the defect on any rejected slate to avoid its inadvertent use by someone else.

Many expensive manufactured fittings such as toilets, basins, sinks, baths, taps, cooking and heating stoves can often be bought for next to nothing second-hand. Even frustrating struggles with

171

Second-hand materials can often be richer in texture than their new equivalents. They are normally much cheaper, and of course they help to reduce the energy and pollution load due to the building materials industry. The illustration shows a kitchen counter made with the offcuts of roofing slates which are normally thrown away.

172

seized connections or imperial fittings, sometimes requiring a professional plumber, do not negate the considerable savings.

One has to be careful, though, that the substantial savings involved are not bought with needlessly prolonged work. I have wasted hours taking apart trussed rafters or removing every nail from wood so it can be planed.

Some expensive old materials are chosen because they are more attractive. Old Cotswold stone roofing, for example, commands fabulous prices. Most materials however are chosen because of financial stringency, yet with careful use, patching and finishing, the results are often better than if new materials had been used. Old wood when sanded and oiled outclasses new in the richness of its grain texture – so much so that there are even techniques of giving an 'antique finish' to new wood!

In much out-of-sight work the savings with reused materials is great, the extra time not too penal. For work that is visible the economic factors are more balanced, but financial circumstances often allow no alternative but to use them. If we just think of them as cheap inferior alternatives to new materials, our buildings will look like just that – and it will have been a lot of extra work too! We can use them so as to benefit by the life they bring – an old floor, an old roof is so much more alive than one of new, industrially processed material. We can build mellowness into our buildings – and it will have been cheaper too.

Materials worth looking for second-hand

RSJs (steel beams)
structural timbers
boards – for: floors (but very time-consuming to finish)
              cupboard doors
              shelving, etc.
floor tiles
slates
roofing tiles
hand-made bricks (for appearance only and if the mortar can be cleaned off)
armoured cable
baths, sinks, basins, toilets, taps, etc.

---

ornamental items
copper cylinders if not damaged or in acidic water areas
solid-fuel heating appliances

Beware of:
pressed steel radiators — life probably limited!
demolition hardcore (with wood or other organic matter in?)
electric wiring (invisibly damaged?)
timber infected with dry rot (or other rot or insects)
timber which has been treated with biocides

17

Organizing information for volunteers

UNLESS THE designer is also the builder (as I sometimes am) there must come a time when information is passed from one to the other. A conventional building specification makes boring reading. A contractor has to read it; few volunteers will! Why provide any more than the barest instructions such as 'Do this like this'? Having experienced work in situations where I was expected to carry out tasks in such an atomistic way with no vision of what I was helping towards, my morale was at rock bottom. I only stayed in the job for the wages. Volunteers cannot be compensated with wages.

Without a sense of the overall picture every little task is deprived of context and the work impoverished in meaning. I have found that showing newcomers drawings, particularly perspective impressions, of the project as a whole, can transform their good-will to enthusiasm. When I am going to be away from the site, I find it is worth going over jobs in detail. When someone has not done something before, much that is obvious to a tradesman is not obvious to them at all. We need to discuss general matters such as the level of accuracy acceptable, in turn implying the tools best suited to the job, such as bow saw or panel saw, for instance; even such minutiae as which nails to use where, at what angles, and whether to drill timber to avoid splitting. Many things can be demonstrated. Slates for instance can be set out on the ground. If paper instructions are necessary – and unfortunately the more

175

remote the architect is from the site, the more needs to be on paper
– annotated drawings are much easier to understand than either
words on their own or 'euro-instruction'-type pictures unsupported by description.

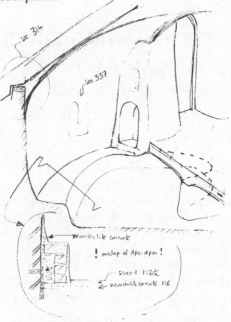

Construction detail sketch for Nant-y-Cwm kindergarten (numbers relate to
further detail drawings).

In describing *how* to do something, I like to state *why*. Often
particular decisions are made on the drawing board, simply
because some decision must be made, when in fact a range of
possibilities is equally acceptable. Tradesmen often know things
architects don't. Even though volunteers are mostly inexperienced, room should be left open should they know better ways to
do things.

Conventionally, specifications are divided into sections for each
trade: bricklaying, carpentry, roofing, and so on. This may make
sense for the building industry, particularly when projects are built

by a system of subcontracts. Of course there are the familiar problems of plumbers cutting through brickwork and letting the spoil bridge cavities, carpenters nailing through plumbers' pipes, electricians cutting the carpenter's structural timbers, and so on, in cyclical sequence.

The levels at which rafters bear on the walls would be nightmarish to calculate, yet their exact positions are defined by the meeting of a conical roof structure with a wall which cannot be located on plan. The curves of the roof arise at the marriage of the two trades (blockwork and carpentry).

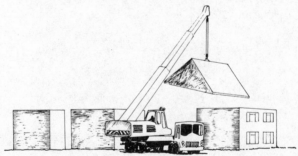

The more orthodox fragmentation of tasks by trade leads to inevitable solutions.

Such fragmented descriptions make no sense for self-builders or volunteers whose concern is not limited to the boundaries of any one trade. They want to see the completion of the whole project. It is better to describe whole tasks. For example, 'Set out rafters on ring beam. Mark and cut birdsmouths over inner leaf of wall below. Nail rafters at top ends and build up blockwork below to support lower

177

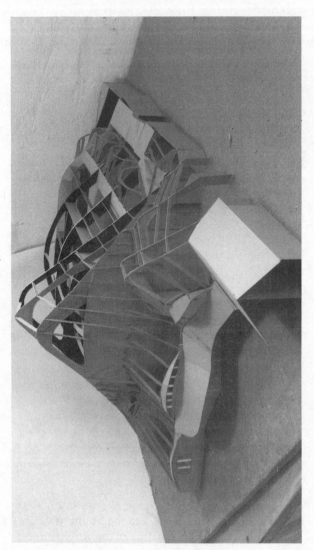

A constructional model not only shows how complex forms can be built but also how these forms can be enriched and woven together. Greater richness of form – and constructional complexity – immediately suggests itself once one starts to make the model. It makes obvious that which can hardly be conceived on paper. The model is a rehearsal for building. Information about how to construct a complicated roof like this, and how to 'find' the form on site, merely need record the process gone through during model making.

178

ends.' A conventional specification would separate the two trades and thus lose the total picture.

Although I try to give a picture of the whole, what instructions are actually about is the step-by-step *process*. For efficiency this needs to be in the right sequence. Various factors influence the order in which jobs must be done. Not all constructional sequences are obvious. Perhaps floor joists may not be laid on the walls until these have been braced by partitions or beams. Numerous little constructional processes can only be done in one order, like slate rivets that need to be positioned before the slate is nailed. Certain forms cannot be described – or located – except by sequential instructions. For example:

1. Raise curved rear wall to ceiling joist bearing level. Leave pockets for easy fit for ceiling joists (every third pocket to be wide enough for frame-anchors).
2. Next: continue brickwork to rafter bearing level for rafters 10-15, remaining brickwork to be stepped down as level of rafters 16 onwards not known.
3. Next: cut and fit rafters 10-15 (birdsmouthed, dpc-sheathed, anchored, etc.).
4. Next: prop into position parallel but 50mm above final level: rafters 16 onwards. Lower ends rest on plate, upper at present in air. Sight across rafter 15 to check that they are parallel. Cut 50mm

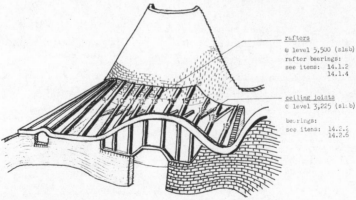

rafters
@ level 5,500 (slab)
rafter bearings:
see items: 14.1.2
                   14.1.4

ceiling joists
@ level 3,225 (slab)
bearings:
see items: 14.2.2
                   14.2.6

Describing how to shape complex forms (numbers relate to further detail drawings).

birdsmouth directly above supports both ends, then place in final positions.

5. Next: fix lower ends of rafters, prop upper ends in position and raise brickwork to support birdsmouths. Build in (details as above).

6. Next: build in ceiling joists, bolting outer ends to rafters. Wait 48 hours for mortar to harden before laying scaffold boards on joists, for access to continue brickwork.

This describes a complex roof form. People have queried how it could be built, let alone shaped accurately. I can think of no other way I could describe this form. In fact every built form arises out of a sequence of operations – this is the sequence I went through in designing the roof by modelling it. The instructions therefore re-

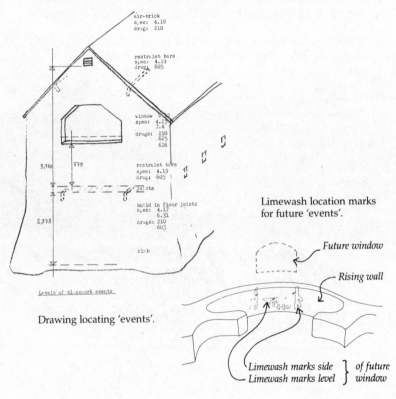

Limewash location marks for future 'events'.

Future window

Rising wall

Drawing locating 'events'.

Limewash marks side } *of future*
Limewash marks level } *window*

180

create this process on site. It needs however some patience to read such long-winded instructions! I avoid them if I can.

Sometimes simple diagrams can show where and when various events occur. On site anything that needs attention, like where, some courses of block higher up, window openings will be, can be marked with limewash to locate plan position and a datum from which to measure upwards.

Access is a point frequently overlooked by the inexperienced. For example this roof (Fig 9) can only be slated in the sequence shown, otherwise how can you position even a roof ladder?

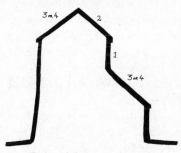

Order of slating roofs so as to retain access to the last bit.

Every job, conventional or not, has a critical path, certain critical items that must be completed by certain dates or in a certain order for the job not to be held up. In most jobs I try, early on, to identify such sequences and paths and those stages when, with future users, things should be reappraised. This can make clear which decisions are best deferred till when. These form the skeleton, upon which can be built a programme of work.

It is not uncommon for a site to have several 'tracks' in operation, such as:

1. Indoor and outdoor jobs   (depend on weather and vary from day to day)

2. High- and low-skill jobs   (depend on labour force and/or supervision and vary from day to day)

3. Seasonal jobs   (must be completed before winter)

4. Deadline jobs   (must be completed by specific deadlines)

5. As-soon-as-practical jobs   (usually the most urgent, yet there are no absolute constraints so they have to be woven in amongst the other factors)

181

Recognizing the principles which shape the progression of the work is valuable not only for the building team but also for the people they are building for, as it can help explain why they don't get what they want immediately!

Much information can be conveyed verbally and informally, but as the job becomes larger and more complex, so does the need for paperwork. Almost certainly, too much paper will be ignored and probably resented. Too little will, in retrospect, have been found to be inadequate. For small jobs, or those under a competent eye paperwork need only be sketches of non-standard details. The next step is annotated sketch instructions given to the individuals who will be doing various jobs, with duplicates kept on file. People tend to lose their instructions right away, and then borrow them from the file on site so it is a good idea to keep a further copy in a safe place. A triplicate carbon job book is handy for recording on-site decisions, though I personally find it hard to draw with a ball-point pen.

From time to time I make a periodic job-descriptions list, listing everything but illustrating and describing in greater detail those jobs that are non-standard. These lists aim to look far enough into the future to keep everyone supplied with materials. There may need to be notes as to tools needed if these must be hired or borrowed.

Jobs with semi-independent teams benefit from a site diary. How often I have explained, demonstrated, and sketched a job with someone who has started work, left at the end of the day and never reappeared on site! Which of the necessary but now invisible tasks have been completed? Have materials already been ordered, or put somewhere for this job? Was there a reason for doing this like this? What still needs doing? Some of these answers can be retrieved from the site diary, but there needs to be some chivvying to get it filled in! In general, I am not a paper-record person, but there are times I wish I had kept a clearer record of who did what, why something was done one way or another, or what should be done when work was recommended after being deflected elsewhere for a while. I like to talk over what to do next, how best to do it and why one way is preferable to another. One-way information fixes things. Conversations on site not only tend to produce better solutions, but bring life into the process by which ideas become a building.

18

Quality control

ALL BUILDING sites suffer from quality control problems. On conventional sites, not everyone understands or cares why something should be done in a particular way. On a volunteer site, the skill and understanding may be lower, but the level of concern higher. On all sites work gets covered up too fast for mistakes to be noticed in time. A building inspector once observed that workmanship is generally better on self-build than on contract sites, as there is every incentive to do a job well, whereas it is not unknown for contractors to bodge and cover up work in the interests of speed and profitability.

Difficult as it is to notice mistakes in time, the real problem is what to do about them. How bad is 'not good enough'? On a contract site it is easy to resort to an established standard. On a volunteer site this would upset people. If I told a self-build client something wouldn't do, he would tell me to go home! There are sure to be mistakes but it is no light matter to tell someone to take down work he or she has just given, to spurn that gift. What can be done? Can errors, without excessive cost in performance, durability, appearance or expense, be satisfactorily rectified? If so, how, and by whom? The same person or another? If not, can the alterations be done discreetly? If work must be rebuilt, how can one delicately discuss the situation without causing hurt? Just because things look badly built are they in fact unsatisfactory?

Blockwork out of level, for instance, may look sloppy but its structural integrity only comes into question when it is out of plumb or badly bonded. I once had to deal with a section of blockwork wall 'plumbed' by the technique of looking down it for straightness – thus compounding any errors! Perhaps unwisely, I decided that it was sufficiently short to be held up by the flanking walls and that the error in plumb had not yet reached a critical proportion. With the wisdom and worry of hindsight, I look at it every time I pass. Fortunately there are, after seven years, no signs of even hairline cracks!

Mistakes in setting out bonding – are they too serious to rectify by using cavity ties or expanded metal lathing in lieu of proper bonding? Can mistakes causing water penetration be protected by a second defence? Do they also have structural implications, like undesirable slipjoints? Which finishes are crucial for durability and which for aesthetics?

The architect is aware of all these issues during the process of the design of the building – but buildings are so complicated in their performance requirements that many are no longer apparent in the finalized design.

As far as possible I try to pre-empt mistakes by explaining why we do something in a particular way: why we try to distribute point loads, why we go to the inconvenience to cut insulation blocks when a brick would fit, why we separate elements to reduce structure-borne transmission, why continuity of ventilation in roof spaces is essential and so on.

In addition to the sorts of things builders often miss, short-cut or ignore, there are all those routine good practices which are not perhaps obvious to the inexperienced, such as supporting dpc trays, or ensuring slopes towards the outside wherever anything crosses a cavity in an external wall. With experience you can anticipate many mistakes but never all.

I try to show people what to do in an educational rather than a critical way and only worry about correcting the essential aspects. Sometimes there are the people who are unaware of the implications of what they do, perhaps 'straightening up' a wall that started out curved. Here, I try to help them to *see* what is happening, to become aware, to cultivate their aesthetic sensibilities. Sometimes we discuss how something could look best, sometimes consider the effect of one or another way of doing something, sometimes just stop, stand back and look.

I do not see quality control as the task of enforcing predetermined standards, nor as the establishment of the minimal acceptable standard – as I understand is common in industry. Rather I see it as the process by which people learn first to do something that they previously could not, then, through this experience, to delight in what they are creating.

If you make something of value, you want it to last. To do that, it must be well-made. For the traditional craftsman, pride in the quality of his work led to his regarding it as almost art. For the volunteer and self-builder, commitment to the artistic quality leads to a desire to make things well, to 'do a good job'.

Supervision then has this particular task: to inspire and nurture the aesthetic aspect – both with a vision of the whole and with concern for the parts; to show how something should be done, explain why this way and not another, and to ease the job around any mistakes that have occurred without loss of performance, durability or quality.

Whereas on the conventional building site supervision is seen as the task of checking up on other people's work and the rigid insistence that they conform to previously specified instructions, it should on a volunteer site inspire, enthuse and assist. It should be a key, not a lock.

Mistakes – how to assess them and make them good		
Is it unstable?	What would make it stable (e.g. lateral bracing)?	Definitely? Or probably?
Will it leak?	What would stop it leaking (e.g. another layer)?	Or is it just a bodge up?
Will there be a nuisance sound transmission?	What would stop this (e.g. built-in furniture)? Will it indeed be nuisance or not noticeable (e.g. other 'masking' noises)?	Satisfactorily?
Will it be durable?	Is durability critical here or is it a short-life element (e.g. furniture)?	What about the nuisance of replacement?

Will it have a heat leak?	Does it matter much? Could it be stopped (e.g. an over-layer of insulation)?	What are the implications? Cold? Condensation, fuel bills, toxic materials?
Is it a fire hazard?	Is it a real danger (e.g. toxic smoke)? Can it be remedied (e.g. buried in sand)?	Is it really safe?
Is it toxic?	Can it be ventilated (e.g. if in a roof void)? Is it in a critical zone or not (e.g. a handrail, constantly touched or an external store)?	Is it morally acceptable?
Will it not perform satisfactorily?	Is it tolerable (e.g. long run-off for hot water)?	Will it be a continual source of irritation, dissatisfaction, maintenance problems, running costs?
Is it unsafe?	Can the risks be controlled (e.g. no children allowed here)?	Is it morally acceptable?
Does it contravene any regulations?	Do the regulations have any meaning in this particular instance (e.g. half-hour fire-resistance of internal balconies)?	Is there a good reason for the regulation? Will we be caught?
Does it not look/ feel right?	Is this just a personal preference (e.g. I don't like red)?	Is it meaningfully wrong?

What then is the reason that it is unacceptable? Is it really responsible to accept it?

19

Management of materials on site

EVERYBODY IS demoralized if work is held up by lack of materials, so I always try to order well in advance. With volunteer labour, however, the speed and sequence of work depend upon too many unknown factors to be accurately timetabled. So materials often arrive long before needed. Unless properly looked after, there will be a lot of waste.

All materials tend to suffer in storage on site, even those that seem so durable such as sand or slates. Wind and rain can carry away tons of sand during the course of a job – and it is not cheap! A client of mine scraped up sand dispersed over his site. For the cost of an hour's JCB time (£7) he saved a lorry-load of sand (nearly £100), not building quality, but quite good enough for bedding paving-stones. Properly boxed in – with old corrugated iron, for instance, very little sand should be wasted. It may however become useless for building if it gets stones, mud, twigs, leaves, etc. in it. If there is a risk, the heap should be covered.

Slates can last two centuries on a roof, but in a stack where rain and frost can penetrate the end grain, one winter can cause damage. Poorly stacked, a lot can be broken – a completely needless waste! Plasterboard is strong enough on the walls or ceiling, but if stored on a damp floor, leaning and sagging, or in stacks which people walk over, it can be thrown away. More recognizably perishable materials such as plaster and cement are, of course, the most vulnerable. As soon as they are delivered I like to see them put in plastic sacks, tied, labelled, dated and stored on wooden scraps in a dry place. Labelling avoids someone looking for, say, cement, opening and not retying a bag containing plaster –in a few weeks it will be useless. Dating prevents older stock getting becalmed at the bottom of the stack.

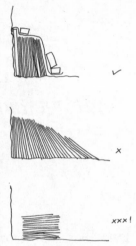

Although timber is usually ordered for specific purposes it may arrive in unrecognizable forms. Truss assemblies or shed panels for instance do not look like 4″×2″s and floorboarding. Ready availability means that it is tempting to use for any old job, sometimes even being cut into pieces for temporary props, scaffolding, short-length requirements and so on.

Stacking slates.

I have found it worth organizing timber properly. Where only a few distinct sizes are on site, separate stacks may suffice. Stacks of varied sizes, however well arranged, eventually get turned over once too often for any order to remain so a rack has great advantages. From one end the sectional sizes are visible, and if all wood is pushed up against an end stop the lengths can be easily seen. It takes time to make a rack so it's only worth it if the job will go on for a long time. Often a stack will do until the roof is on and thereafter timber can be kept indoors.

Timber needs protecting from rain and from moisture from the ground. Wet wood is much harder to saw, in accuracy as well as effort, and it is not long before such timber suffers the first stages of fungal decay. Polythene covers tend to hold huge pools of water. Unless laid crossways, corrugated iron just directs water into the stack. This takes up more space so you need quite a bit of room around the stack if it is to be kept in good condition. If it can be positioned at the place where the timber will be needed and where it can be unloaded directly off the lorry, you can save a lot of heavy, time-consuming and disruptive double handling. If you buy largish orders of new wood it can be mechanically unloaded off the side of the lorry. Small or second-hand orders are usually pulled out from the back. Lorries like hard ground; some drivers refuse to drive on to the site if there is any risk that they can't get off, and will hold you liable if they get stuck.

On a commercial site, short pieces of wood are too much nuisance to have around so are thrown away. But buildings use a lot of short pieces – noggins* for instance are usually 14″ long. Crudely sorted offcut piles can save cutting a few inches off long lengths making them too short to use. Similarly, stacks of broken blocks and of shuttering grade planks can economize on using new materials. All these sorts of things need to be placed for easy accessibility but where they will not clutter the site.

The appropriate balance between thrift and tidiness depends on the situation. To a large extent their inherent conflict can be resolved by really organized storage. Stores need to be in the right places, otherwise additional piles will start to grow up elsewhere.

There is always the risk that high-quality material is damaged by being used for low-quality tasks. Joinery-grade timber can get used for shuttering or barrow planks, floor tiles to level the cement mixer, or slates to wedge scaffolding. I try to keep a stock of low-grade materials for such functions, in separate unprotected stacks so that it is obvious what they are. If need be there can be a sign.

Wastage is often as high as 20 per cent on conventional building sites, where there is little incentive for care and thrift. Waste in fact is often cheaper than care! While I may order 20 per cent over for some things like slates, I aim to keep wastage due to our work well below 5 per cent – and in any case many of the offcuts, conventionally just thrown off the roof, are used for cut slates or even window cills and splashbacks around basins. Wastage is expensive in monetary terms but if a reluctance to throw things away leads to clutter and thence chaos, the price is in morale. Projects may struggle for lack of money but they can die for lack of morale. Small items suffer most from chaos. The more things that cannot be found, the more others are rummaged through and disorganized. Spending fruitless hours failing to find things in a

---

* Pieces of wood wedged between joists, rafters or studs.

chaotic jumble of stuff, often in poor light, is thoroughly demoralizing.

Experienced self-builders build the garage first. Not to have a roof to live under, as was my first priority, but to store tools, materials and perhaps even have some workshop space. This, or any site hut, must be well laid out, and well provided with shelves, even if only planks on bricks. Piles of stuff on the floor can never be kept in order! If you will be working in winter, it needs a lot of light. Roof lights can't be obstructed as can windows. Electric light is a fantastic asset. You can see to work outside until well after sunset, but on a cloudy winter's day it can be too dark to find things indoors.

When converting old buildings with insufficient money, space or time to build a hut, one room can be used as a storeroom. This is where problems start! The slower the progress the higher the pressure to occupy semi-finished parts of the building – and this may well include the storeroom. Stores may be moved three or four times, losing any semblence of order at each move. Small things such as soot-doors, door-closers, locks and screws easily get lost. It hurts to buy expensive items that I know are on site, but just cannot be found. The price in wasted time and materials and general demoralization has led me to realize that a storage shed is cheaper than it may at first appear. You don't have to buy one. They can be built quite cheaply from scraps but this takes time. Here you need to consider whether morale is better served by a purchased hut which allows you to start actual work sooner. Unlike amateurs, professional builders rarely get straight to work. They get everything ready first, levelling for barrow paths and cement mixer and so on. The time taken to sort out the site before you start doesn't take long to pay for itself.

To keep some degree of control over where things are, I like to have the storeroom well supplied with strong cardboard boxes, paint tins and beheaded gallon cans, all boldly labelled. Boxes are for types of goods: electrical, plumbing, rain-water goods, boxes of screws, and so on; the cans are for nails, bolts, etc. I like to have a surplus of such receptacles to avoid someone tidying a random collection of oddments into a specifically marked box, safety helmet or bucket.

We must all have experienced jobs which cannot be started till something in the way is moved, and this cannot be moved till there is somewhere to put it and that somewhere depends upon the first job – and so on, cyclically! The 'right place' for things, however, is not always so obvious. Where something must be moved, for instance tipped blocks moved out of the way, I try to ensure that their movement is part of a flow across the site to their final destination. I try also to arrange that as much as possible travels by barrow rather than having to be carried, and that it travels on the flat or downhill. I also try to ensure that every task has enough space and has all necessary materials immediately to hand – for brick-laying, stacks of bricks about five feet apart and space to manoeuvre a barrow of mortar; for carpentry, space to lay out tools and to cut timber without having to move elsewhere. I like robust saw-horses and enough space so that two people can cut both ends of timbers simultaneously, otherwise one waits while the other saws. A workyard for power tools accessible to all who are using them avoids the need to wander around the site looking for an electric lead!

When there is a high turnover of people it is hard to keep things in the right place. Tool shadow boards, labelled shelving and even writing where things belong on them, can help. Basically, however, it is a matter of people. If a regular core group, however small, can start first and finish last, or, even better, if the whole workforce can come together at the start and close of the working day, it is much easier to keep control over what is where, see that everything that should be is returned to its correct place, make notes as to uncompleted jobs, materials that need ordering, clean up and generally keep up a good atmosphere. Even if people come and go in their own time, opportunities can be found to form a rhythmical measure to the working day – formalized and social lunch and tea breaks for example, especially where everyone can sit around one table. This can bring a sense of unified purpose to otherwise disparate tasks.

It is these mechanisms of order, unity and purpose that underlie good housekeeping on site, and do so much to make the workplace and the work enjoyable instead of exhausting. Good housekeeping is also good economy – much money can be wasted on double ordering, spoilt or misused materials, and much priceless energy can be squandered in working in a mess, endlessly clearing space to work, stepping over things, searching out materials. Inefficiency and disorganization can double the amount of work needed!

By choosing to use second-hand materials and thereby accepting the time-consuming complications involved, the limiting factor becomes less that of *money*, than of *will*. However, in creating and maintaining the necessary good atmosphere on site –the essential foundation for productive efficiency – will alone is not enough. Will needs to be constantly nurtured. To achieve this, the site must not be squalid but attractive; work must not be needless but visibly productive; the project as a whole must be beyond the personal or utilitarian, and must be inspiring.

Communal will has the strength and ability to overcome, and the imagination to sidestep, apparently insuperable financial problems. Financial problems are in fact ultimately problems of communal will. Likewise, problems of will are ultimately problems of inspiration. If the inspiration is clearly, strongly and socially held, the necessary will has already come into being. Will without clear and shared inspiration is like building on dubious foundations: much can be achieved, but it is not sustainable.

Checklist
Is there on site:
   A felt-tip marker?
   Surplus boxes?
   Surplus tins?
   Covering for timber stacks?
   A shadow board for tools?
   A notepad for orders?
   A rubbish zone, bag, box, or skip?
   Plastic bags for cement, plaster?
   Adhesive tape or string to secure them tightly?
   Barrow planks?
   A broom?

20

Safety on site

ACCORDING TO statistics, building is, after mining and trawling, the most dangerous of industries. It is three and half times more dangerous than the industrial average. In a working lifetime, the average building worker can expect three or four accidents that will keep him off work for at least three days. One in fifty will die in an accident.* These figures take no account of long-term damage to health.

One common view of accidents is that a high proportion are due to carelessness amongst the workforce. Another is that management is more concerned with profit than safety and hence allows environments where risks are enhanced. From the point of view of the self-builder or volunteer, which party is to blame is purely academic. The important issue is accident prevention. Ignorance and disregard of hazard prevail in the building trade. Although it can be complicated to insure non-employed volunteers, insurance can protect the organizers from the possibly crippling financial implications of an accident. It cannot protect the victim from the physically crippling effects. In concern for inexperienced volunteers one cannot be too careful.

It is in the self-employed person's workshop that machines,

* Patrick Kinnersley, *The Hazards of Work: How to Fight them*, Pluto Press, 1973.

unguarded as in Victorian times, are to be found. Likewise on volunteer and self-build sites, safety measures are likely to be more constrained by cost than elsewhere in the industry and there is no union representative to counter this. The workforce may not be careless, merely inexperienced – it amounts to much the same thing. The risks, in other words, are high. Hindsight invariably shows how something could have been avoided. We must try to replace hindsight with foresight.

This chapter is an attempt to highlight common hazards so that dangerous situations can be identified early enough.

Even on the smallest building project, there are too many heavy objects, dangerous materials, sharp or mechanical tools, high-level working positions and improvised or abused electrical equipment, for accident risk to be taken lightly. In addition to this, the modern building site sometimes seems to be a Pandora's box of nasty substances. It is! Some can be avoided, others used more sparingly, but all deserve appropriate precautions.

On top of the building industry's poor record for safety, must be added the fact that unfamiliarity with jobs, tools and the work site increases risks. For this reason, the first week is the most hazardous. Unlike regular builders, volunteers come and go and there are also some whose risk-consciousness never rises above the first-week level.

Risks occur from the conditions in which work is undertaken, the tools and equipment used and the building materials involved. They can be broadly classified into risks of immediate injury, and risks to health in the longer term. Some risks are obvious, so I will concentrate on those which, to the sort of people I have met on volunteer sites, are not:

Common causes of accidents and injuries	
Falls, due to:	ladders moving
	unstable scaffolding
	inadequate or cantilevered scaffold planks
	awkward working, moving or
	carrying positions at high levels

196

Blows, wounds, and other injuries	dropped objects
	projections
	sharp edges
	protruding nails
	injury by machinery
	electric shock
	chemicals or particles in the eye
Long-term damage to health	residual effects of injuries
	strains
	toxic substances

There is one inviolable rule when working at high level. Never be on your own on site. If you fall, you could lie on the ground for days until somebody found you – and one cold night on top of the shock of a serious injury might well be enough . . . .

Ladders cause a lot of accidents. They should be tied at the top and blocked at the bottom. But often there isn't anything to tie to at the top, and the bottom is uneven ground so that it is hard to plant the feet firmly without little chocks which always seem to vibrate out of place when you are halfway up. If the ladder is at all insecure, someone else should hold it.

The main risk from ladders comes from working off them. It is often tempting not to bother with scaffolding and just work from a ladder but too often this work extends to the side far enough to require one to lean over, and then . . . the ladder slips. Even used for access, there are risks. If the ladder is too long, the part protruding above platform level can counterbalance the lower section enough so that when the inexperienced press the top bit, the bottom section slides (Fig. 16). Substantial blocking can prevent this.

For access to roofs, the ladder may not rest on the gutter without damaging it. Consequently, unless you have an offset frame, it is a rather awkward step up, and even worse, back down. As I don't feel comfortable just stepping hopefully into thin air, I chalk the ladder position on the roof.

People tend to move ladders that are in their way, propping them in positions that are in no way safe, yet superficially may appear so. If all ladders are routinely tied or propped, even when apparently unnecessary, it is easier to distinguish a casually parked from a safely positioned one. A bit of cord tied to the ladder increases the likelihood that people actually will tie it!

Moving goods up from platform to platform *inside* a scaffold tower.

Brooms, and so forth, should *never* be leant against ladders or scaffold. They can catch in trousers and trip people, or at worst go up the trouser leg in a fall or innocent jump – and cause the most appalling injuries.

Unless you are familiar with it, conventional tube scaffolding is complicated and time-consuming to erect, and dangerous if any of the heavy unwieldy bits get dropped. For most self-build and volunteer sites, framework towers suffice. The very cheapest are too lightweight for building purposes, but the next grade are cheap and adequately strong, although prone to rusting at the welds. They are designed to be climbed on the *inside*. I have learnt to beware of oily spills such as creosote which can make everything much more slippery than expected!

Towers that are set on bricks or small levelling pieces, or are not firm and level, are prone to walking around with the risk that one leg will come off its base. If tall, towers should be strutted or tied to the building. If, as sometimes happens, the diagonal braces are forgotten the tower can distort and collapse.

It is tempting to overspan or overload scaffold planking. The first failure is twisting – which can throw someone off if the platform is not wide

198

enough. The next is breakage. I have seen both, fortunately near the ground.

Nails in their undersides stop planks sliding off the frame. Cantilevered planks should generally be avoided, and should always be tied at their other end to an anchorage well down the tower, not just to the last piece of scaffold. Even if you know of their existence, unsecured cantilevers are always a risk as you cannot see out of the corner of your eye exactly where to tread.

It is frequently convenient to build in joists as work rises and use these as a working platform. Often a temporary floor of unsorted, uncut and unfixed boards is laid on the joists. Random board lengths tend to result in some projecting far beyond their supportive joists. It *looks* like a safe floor, but when you tread on one of these board ends . . . .

Careful layout can avoid, or protectively cover, such cantilevers, but from time to time people use or move boards, for instance to level up for a sawhorse, to get access to services below or make space to saw between the joists – and the cantilevered board ends become unprotected. Such a floor needs constant attention to keep it safe. To make matters worse, sections may be covered up with polythene, plasterboard, scraps or debris – and who knows what is beneath?

For most falls it is what you land on that counts! That was certainly my experience when I fell off a roof, landing on a stone pile. I now always try to keep the area below where anyone is working free from nasty things to fall on!

On a frosty morning, scaffold boards can be slippery. Grit and dust have a similar effect. It's a good idea to sweep platforms clean from time to time. Likewise, I like to sweep off the debris of cut slate that accumulates on roofs. It is too easy to dislodge a perhaps small, but sharp bit on someone below.

Bits of bricks, wood and even tools are often inadvertently kicked or knocked off scaffolding. Anything balanced, badly stacked or any debris on a high level is dangerous. Having knocked a parked hammer onto somebody's head and in my turn, been hit by a falling plank, I know! Furthermore, if you are working in a precarious position and lose your balance, the instinctive reaction is to drop what you are holding and grab for support! Bad luck for anyone below!

I find a great resistance to the wearing of safety helmets and few people even wear steel toe caps. With or without such protection, the golden rule is never to work above (or below) someone else. During passing-up or throwing-up operations the rhythm of work should be such that the person below can keep an eye on the person above and only move 'blind' when it is safe to do so. I learnt that by once nearly dropping a concrete block on someone's head when its corner crumbled in my hand. I still don't know how I avoided killing him.

The person above should never need to shift his feet or move out of balance, and should be able to drop a misthrown brick, scaffold coupler or whatever, in full confidence that the person below is aware of what is happening. The alternative is to risk falling in attempting to catch it! Things passed or thrown up do get dropped from time to time, as do heavy objects that are being moved, lifted or 'walked'.* I like to have a clear space beside or behind me to jump into – and I've needed it more than once!

For anything so heavy that I may lose my grip, or that might fall apart, I always stand with my weight on inward pointing toes, so that I can bounce it off the thighs. Occasionally, concrete blocks have hairline cracks. If one breaks in the hands there is one half for each foot. It has happened to me – but I keep my toes in! Sometimes really heavy objects can be lifted against a wall; you can hold it against the wall if it starts to slip.

We take minor hand injuries for granted, but with heavy weights, sharp edges or machines they may not be so minor. Industrial compensation rates thumbs at many times the value of fingers. Moreover, they stick out first and so are at greater risk. I keep the thumbs in, when moving heavy objects, barrowing through narrow openings, sawing and so on.

Projections such as scaffold poles at eye-level are small to see and easy to walk into. They should be padded or wrapped with rag 'flags'. Nails in door jambs, expanded metal lathing protruding out

---

* *Walking* (a heavy object) Holding it upright and moving one corner then the other whilst keeping it in balance, so as to move it across the floor.

of walls, and so on, wait to catch anyone passing. If they cannot be removed I fold them in.

Treading on nails protruding through wood is a common and wholly unnecessary form of accident. You can also rip your hand unpleasantly when handling such timber. I don't like old nails. I always knock them through or clench them over in the course of demolition, deshuttering, or at any other time I find them. Deep wounds such as these cause need medical attention due to the risk of tetanus.

Eye injuries are all too frequent. It is sensible to wear an eyemask, or at the very least a low-brimmed hat when chipping or disturbing grit overhead. An eye-mask is also needed for work at small openings or chimney-effect ducts, where grit is raised by the draughts that pass through them. As a result of unpleasant lessons, I have learnt to close my eyes and brush off my eyebrows after sawdusty or gritty work, and to close my eyes at the moment of impact when cutting bricks. Anything in the eye hurts, but to use power grinders and the like without eye protection is to risk blindness.

Splashes in the eye from even relatively innocuous chemicals such as industrial detergent are unpleasant and can cause lasting damage. Building sites, however, abound with much more dangerous chemicals, notably wood preservatives. There is hardly a can on site which does not include in its instructions 'avoid splashes to eyes, flush with water immediately'. Who, perched high on a scaffold, blinded and in pain, knows where the nearest water is?

I recommend two simple rules when working with chemicals:
1. Never work above your head without eye protection.
2. Always have a bucket of clean water accessible.

If you apply for employers' liability insurance for building work, there are questions on the form about what machinery will be used on site. And no wonder, because any accident you can have with a handtool is going to be many times worse with a power tool. In spite of the fact that chainsaws have safety guards and chain stops, they cannot be relied upon to always work, any more than can the foolproof rules that, if followed, would make axes 100 per cent safe. I know enough people who have witnessed horrific accidents with

201

circular, chain, and band saws, routers and planers, overturned machinery, snapped cables and all sorts of agricultural machinery, to be wary of machines. The worst I have seen was a finger caught in a hydraulic crane grab, and that was bad enough. Hand-held mechanical saws are made to be held with two hands. One alone cannot guarantee it goes straight enough not to jam in the cut and jump back at you, probably across your work-holding hand.

Big machines are big; I keep well clear of them. Even small ones are so much more powerful than human power – and they don't know when to stop. Machinery greatly eases work. Every machine is a willing servant until something happens, then suddenly it is a monster. Machinery needs to be used with care, and by people who know what they are doing.

If you adjust electrical power tools while connected, you risk them starting by accident. They should always be disconnected when not in use. Electrical equipment generally must always be considered a risk in the sorts of conditions prevalent on building sites – sharp edges, and foot and wheeled traffic which can damage cable, together with generally damp conditions. Regulations require all outdoor electrical tools to be protected by a current-operated trip switch (ELCB). Every year there are hundreds of fatalities, largely from garden and do-it-yourself equipment that such protection would have prevented. Having had – fortunately minor – shocks, I believe in such protection for *all* electrical equipment.

A knowledge of basic resuscitation technique – the same as for drowning – can save life. There may be secondary injuries caused by falling, which can be more serious than the original shock. Every site, however small, needs a first-aid kit, accessible at all times – in other words, not locked in another part of the building, or in the foreman's car which may be on an errand somewhere! But does everyone know where it is? When I last needed it in a big hurry, it had been moved to a new cupboard.

It is a good idea to have at least a rudimentary knowledge of first aid, access to a telephone and know where the nearest hospital is. If someone can ring to forewarn them of a casualty on the way, it can speed up treatment. In the crane accident I described, the road was blocked one way by a lorry, the other by a herd of cattle – but some sites have no car, or may be quite remote. It is as well to think about how to summon, or get to, help in an emergency.

Accidents are dramatic and highly visible. Long-term damage to health is not, but it is no less serious. Back pains are one of the most common ailments. It is not easy always to bend your knees when picking up or putting down something heavy but if you spread your legs or turn them to the side it is easier. It is also important to keep the back straight; looking up helps achieve this.

Moving blocks to avoid lifting.

Personally I prefer to avoid lifting. I like, for instance, to move blocks from one stack at waist height to another, never running the pile down so low that I have to bend to pick things up. And the bottom blocks are wet and so heavier as well as being more awkward to lift. Heavy things I pick up one end at a time or walk them up a wall or stack until they are at the desired level, so that I am only *lifting* half the weight at a time.

The same approach can be applied to most things that have to be moved. Must it be moved at all? Can it travel by barrow or must it be carried? Loading a wheelbarrow forward means that although the barrow will be less stable when standing, there is less weight to lift.

Work positions to avoid lifting.

It is easier to build a wall from waist-high than ankle-high stacks. If possible I stand in the foundation trench to work until the wall is high enough for it not to be too low again when the trenches are backfilled. Hardcore and concrete slab can also wait until the wall is high enough to need a new level to work from. As they say, 'You are only as old as your back.' My back is old enough for me to seek to avoid working it. Not all lifting can be avoided but a surprising amount can be.

Safety checklist
Where is the first-aid kit?
Does everyone know where it is?
Who maintains it?
Does anyone know first aid?
Are all poison containers and brushes prominently labelled?
Are there nitrile rubber gloves
                              dust masks   on site?
                              eye masks
If second-hand or demolition wood is on site – have the nails been knocked over?
Where is the nearest telephone?
Is there a car available? Where is the nearest casualty department?
Is water to hand when there is a risk of splashes by aggressive chemicals?
Is all electrical equipment in a safe condition and with ELCB protection?
Are working practices safe?
Is there appropriate insurance to cover liability in case of accident?

Only a few generations ago, building materials were limited to timber, straw, local stones and soils and slaked lime. However bad elsewhere, building work itself involved virtually no health risk from the materials used. How different today!

The building site storeroom is a veritable witches' cauldron of poisons. When children get in and mix up 'poison' they really do! Building materials are often treated as though they were as harmless as they were in the past. On how many sites do you see gloves worn to handle impregnated timber? I have seen wood preservative used as a decorative stain – and even been offered it in the builders' merchant for this express purpose! Sometimes the brush handle is floating in a tin of the stuff – or even in a saucepan, perhaps later to be used by children playing. Timber preservatives are highly toxic through the skin, not to mention when it gets eaten with sandwiches.

Modern building materials number some 21,000, made up of perhaps 1,000 chemical substances.* Far too little is known about their effects, even less about their effects in combination. Research suggests that in western society one in three of the population suffers health problems due to indoor pollution.† Building materials play a large part in this. Gases given off by paints, glues, solvents and biocides are stronger in the first few days, so builders get a bigger dose.

The risks to health increase with repeated exposure, but some people are particularly sensitive. I try to minimize usage of chemicals, warn people, and always take precautionary measures. It is as well to avoid concentrating exposure on a few individuals so that for instance it is not always the same volunteer who day after day handles fibreglass. Pregnant or nursing women, or children should not handle chemicals or be exposed to their fumes.

Common building materials and risks associated with them

Material	Harmful characteristics	Precautions
Asbestos	highly carcinogenic	Don't use it. To avoid inhalation, always wear face masks and brush off your overalls out of doors.
Asbestos cement	contains asbestos	As asbestos. Water when cutting and drilling and do so only in the open air. Brush off clothing.

* Arbetarskyddsstyrelsen *et al.*, *Sunda och Sjuka Hus*, Planverket, Stockholm, 1988; Kerstin Fredholm, *Sjuk av Huset*, Brevskolan, Stockholm, 1987.

† Esko Sammaljärvi, *'How to build a healthy house'*, in *Det Sunda Huset*, Statens råd för byggnadsforskning, Stockholm, 1987.

Asphalt and bitumen	risk of burns	Hot processes are best left to specialists.
	carcinogenic	Wear gloves and clean hands immediately after.
Bricks and blocks		Guard eyes when cutting.
Cement	caustic	Avoid undue skin contact.
Dampproof injection	heavy organic vapours from solvent are toxic	Good ventilation, do not work in this area until fumes have cleared.
Debris	can contribute to accidents, e.g. slipping or tripping; if burnt can produce toxic smoke	Clean up.
Dust	can include mineral fibres and toxic dusts	Vacuum and bury. Highly dangerous materials should be dealt with by specialist contractors.
Epoxy cement	toxic through the skin	Always wear nitrile rubber gloves.
Exhaust	toxic, includes carbon monoxide	Site engines (e.g. pumps, mixers) where exhaust does not blow into buildings, trenches, etc.
Glass	vicious if accidentally broken	Handle and store with care. Protect glass in storage, or drape prominently labelled paper over it.
Insulation materials	fibres carcinogenic	Avoid inhalation of fibres, always wear masks. Do not burn plastic materials, e.g. polystyrene.

Lead	toxic by ingestion and through skin	Wash hands after handling. Ventilate very well if soldering or otherwise melting.
Lime	caustic, a very light powder	Avoid undue skin contact and inhalation. Hold your breath when opening sacks or do it in open air.
Oil Paints	slippery, dermatitis risks contain a variety of toxins; potent fumes from polyurethanes and melamines	Clean up. Ensure good ventilation.
Pastes (wall-paper)	some contain mercuric fungicides	Avoid this kind.
Plaster		Avoid inhalation.
Plumbing	toxic vapours from plastic solvents	Do not use solvents in a trench. Ventilate well.
Preservatives (timber)	highly toxic by ingestion, inhalation or through skin; Permethrin and borax-based ones are less toxic	Wear nitrile rubber gloves, overalls. Ensure good ventilation. Use only where necessary, have water to hand to sluice eyes.
Slates and tiles	zinc and lead from nails and flashings	Wash hands afterwards.
Steel	sharp cut edges	Wear leather gloves.
Timber	often preservative treated; Copper-crome-arsenic salts become stable and safer to handle; organic solvent-borne preservatives do not.	Wear nitrile rubber gloves.

| Welding | fire, electrical and fume risks; arc-welding can cause eye damage | Avoid welding painted, galvanized or oily materials, especially in confined vessels or spaces. Ensure no bystanders look at arc flashes. |

Smoking often enhances adverse chemical effects and is a fire risk with inflammable vapours.

VI

CONCLUSION

21

The road into the future

OVER SOME eight years' involvement with volunteer building, I have learnt that necessity and enthusiasm alone are insufficient to cope satisfactorily with the special requirements of this kind of work. In almost every aspect both the needs and the appropriate means of fulfilling them differ from the conventional norm.

While almost all building these days is constrained by cost, gift-work building projects are likely to be based on funding far and away below the level at which building by normal processes can be thought possible. In all probability economy will have to be exceptionally stringent throughout. In gift-work, unlike contract work, however, this does not imply a reduction in quality. In fact high-quality work which is normally too expensive to afford costs nothing.

There is a curious reverse logic here. I have designed projects which, because they have little money, depend upon gift-work. They can therefore afford non-standard craftsmanship which would be too expensive were they 'normal' projects depending upon 'normal' building contracts. For no extra monetary expense, these buildings are more interesting, aesthetically rewarding and therefore inspiring to work on. What, in the conventional world, would be economically impossible is, for gift-work, will-sustaining.

Many qualities that people need in their daily surroundings are too expensive to have built. They are vital for health of soul, but just cannot be afforded – unless we change the ways things are done. Even if self-building is restricted to, say, finishes, we have made a step towards homes healthier for the spirit than just storage boxes for people.

A building with an inviting soul can *only* be built with the attitude of gift. This is possible but harder to inspire with paid employment, where time now has a monetary price. The gift of labour brings both visible and invisible benefits to the quality of work. Visibly it is the difference between a meal cooked with care and pleasure and one executed as an occupational requirement. Invisibly, but tangibly, the values of the builders are imprinted into the environment.

Because labour is free, time can be squandered on needless drudgery, or it can be used, for instance, on hand-made intricacies and texturally rich hand-finishing. This approach allows every aspect of the building to evidence individual attention and care, to have an individual signature and to be impressed with the work of the human hand. And what a difference there is between environments that record their making by hands or by machines!

Freedom from the constraining rigidity of a building contract allows work to develop both human potential inherent in the builders, and the environmental potential which becomes apparent in the unfinished building itself. Neither can be anticipated on paper. Both require that the habit of giving orders be replaced by an attitude of listening. Participation and evolving design are not without their problems – far from it! But the potential inherent in this process outweighs the disadvantages.

The architect has the precarious role of trying to maintain individual contributions and holistic overview in balance. Indeed all the time-honoured hierarchical relationships are unsustainable once they are no longer determined by the relationship between the paymaster and the paid. Relationships based not on taking or buying, but on giving, require reciprocal gifts. One gives labour, the other opportunities for personal development and fulfilment. Whereas the old relationships were those characterized by bonds of duty, the new are characterized by mutual resolves of responsibility.

In this context, the old style of leadership – leaders empowered to command others – is inappropriate and disastrously divisive.

Leadership now is by inspiration, and, as such, it is the group rather than the individual that is its appropriate channel. Projects that are fuelled by the enthusiasm and gift of energy of a group, but have inherited or acquired a leader who *sees himself* as having a mandate of authority, are prone to social strife. Leaders on the other hand whose wisdom and magnetism lead them to be taken as guru-figures can accomplish much without strife. Giving people answers does not foster their inner development but leads to dependency. Such leadership is but a new form of an old way.

For basic reasons of practicality and morale, in gift-work projects so many things have to be done a different way. Even the way the work is undertaken benefits by reorganization into forms appropriate to the volunteer situation. Project planning and management are demanding due to the need for fluid policies to take account of the fact that at different times money, energy, and speed of completion may each in turn become the limiting factor. Inappropriate policies can have financially disastrous results, or can lead to work that is frustratingly slow or perceived as exploitative. Well-attuned policies allow work to maximize invariably scant resources.

While inspiration provides the incentive and nurtures the will, so also the will with which a project is carried forward, visible as the building rises into being, inspires others in ever-widening circles. It is through the actual building work that inspiration can take root and its fruit is inspiration for many others. It is this that unlocks donations which in turn permit the will-energies of the volunteer group to give progressively more substance to their inspiration.

A cycle of development is thus set in motion, fuelled only by the principle of giving. But although growth may be inevitable once it has started, the actual moment of starting is hard. It has nothing to sustain it except faith.

My experience has led to me to the conviction that if the deed is right it is *by definition* practical – however much it may, at first sight, appear not to be so; that if it is right, it will succeed. But while one may blithely rely upon money somehow appearing, it is not advisable to be too confident that the deed is indeed right! The most honest, objective and soul-searching appraisal of a project

213

and its very foundation is not just a moral responsibility owed to those who will give time, energy or money. It is a fundamental practical necessity.

Even if the foundations are indeed right, the growing project can never be guaranteed to be far from trouble. At every turn we can no longer fall back upon established ways of doing things, but must constantly be alive to the demands of the developing situation. Work given, for instance, may not be the appropriate work to receive. How can we genuinely thank the intention, yet not accept the offered gift – for instance, if someone offers to chain-saw a tree that was a central feature of the design? How can participation be encouraged which will contribute to, rather than divert from, the overview? How can the gift of work of inadequate quality be accepted rather than spurned?

Much can be done by anticipating problems in building construction, risks to health and safety, damage, wastage and misuse of materials, failures of communication and co-ordination; by developing a listening rather than assertive approach, and group rather than individual accomplishment. Anticipation is always easy with hindsight – it is never so easy at the time! None the less it is helped by the benefit of experience. In this book I have tried to share my experience so that others need not make the same errors as I have made.

These are, of course, by no means the only problems! Indeed the path of volunteer building is not the easy option. It is harder than relying on conventional precedents. But that is only to be expected in that it is based upon values which stand those of the conventional world on their head.

If we look around us today, we see that although it is actually good-will that holds society together, it is widely supposed to be the principle of gain that does so. Numerous social mechanisms from legislation to commercial relationships have grown up, founded upon this principle of gain to the extent that money has become ever more widely accepted as a common denominator of worth. We live in a culture in which it is widely accepted that a person's work, time – in one sense life – can be exchanged for money. It is not surprising that if the value of a person is viewed in monetary and mechanistic rather than in spiritual and living terms,

214

then that which is spiritual in our environment – that which comes from the realms of inspiration or reverence – is reduced to monetary values.

The society and the environment that result are no accident. On the one hand the path of materialism leads from need to greed, from greed to fear, from fear to threat. Almost half of all British and American scientific work* and more than half of British government research funding† is allegedly arms related. At the same time, in many communities, over 25 per cent are unemployed.‡ Of the remainder, how many enjoy fulfilling work? The bleakness of unemployment – of being unwanted, one's gift unacceptable – is made more acute by material expectations in a consumerist society. This growing lake of useless lives is all the more tragic as in reality it is a pool of untapped energy and talent in a world where there is a crying need for important things to be done.

Leaving school to unemployment, young people learn to take *what the state gives,* but what, even if they do not realize it, they hunger for is the *opportunity to give to society.* This our society denies them. In our economy, service industries grow, but service-giving declines. The employed are bound by mortgages, hire-purchase and job precariousness in chains of dependence; the unemployed are seen as a burden upon them. What a way to look at human life! What a waste of human potential. What a negation of that which is truly human in individuals.

The principle of taking is the shrivelling path of dehumanization, of environmental rape, of social collapse, the war of all against all. It brings us to the threshold of the abyss. The principle of giving on the other hand is the path of healing, of development, of harmony, beauty and peace. The difference is that between money and

* 40 per cent of engineers and a higher proportion of physicists (Sir Martin Ryle, 'Wrong research' *Resurgence,* no. 112, Sept.-Oct. 1985).
† 58 per cent of government R&D funding. In addition to which the Ministry of Defence spends £8 billion a year on some 45,000 defence contracts (Roger Eglin, *Sunday Times,* 22 September 1985).
‡ If you live in the fully employed south-east of England, go west or north!

wealth of spirit. It is no accident that the clouds of violence and ecological calamity gather menacingly about the fringes of society.

If, on the other hand, we recognize that society cannot survive as society where it is not held together by giving, that men and women cannot develop if their spiritual essence is not recognized, appreciated and nurtured, we must not only look at things, but also go about them, in a quite opposite way.

In this light, it is essential that work be meaningful, fulfilling and nourishing. To be meaningful it must be clear that it is filling a need, serving others – that it is in effect, morally good. To be fulfilling it must engage the whole being – head, heart and hands. Through this we can find balance, greatly lacking in the world today. To be nourishing it must be, in a way, artistic. There must be room for wonder, care, reverence, in our daily environment and daily deeds. Without this, how can we hope to be whole? Who could work for months, for years, amidst the ugliness and mechanical brutality of the conventional building site if they were not bought off. Perhaps other people can; I certainly could not.

If work is not experienced as the process of raising transformation – as when the intransigent ugliness of building materials are transformed into an aesthetically nourishing environment – it is not able to support the human spirit. If it is not inspired by a moral ideal, the will is not renewed. In such circumstances work is a drain upon our energy, a burden. It cannot be sustained without personal loss.

Work is love made visible.

And if you cannot work with love but only with distaste, it is better that you should leave your work and sit at the gate of the temple and take alms of those who work with joy.

For if you bake bread with indifference, you bake a bitter bread that feeds but half man's hunger.

And if you grudge the crushing of the grapes, your grudge distils a poison in the wine.

And if you sing though as angels, and love not the singing, you muffle man's ears to the voices of the day and the voices of the night.

Kahlil Gibran, *The Prophet,* Alfred A. Knopf, New York, 1963, p.28.

216

Without a gift-work approach, this building would have been impossible. Not just for financial reasons, but because in every detail, it is the contribution of hand and heart that make the atmosphere – and these cannot be bought on the open market.

217

Work brings us into relationship with matter. It requires effort –both physical effort and effort of will. What distinguishes work that is stimulating and invigorating from work that is exhausting and energy-sapping is whether or not it is inspiring.

Gift-work can offer no cash rewards. If it cannot offer inspiration it cannot last the course. It shows up more clearly the critical issue that is generally obscured in paid work but which society none the less cannot put off facing for much longer. Can work be inspiring? Can it be a joy?

If we look around us, we must ask: Can it afford not to be?

AS WORK needs inspiration to transform it from an exhausting to an invigorating process, so also does inspiration need work to carry it out. In the realm of work, matter and inspiration need each other as much as do service and need.

Inspiration, idea, spiritual values, are mere dreams until they can become united with the world of substance. Matter is the foundation of materialistic values unless it can be raised by artistic means to become imbued with spiritual value. Spiritual values need to be rooted in matter, as matter needs redemption by spiritual values.

Neither social life nor the living earth itself can survive unless we take this path. But it will have to be a conscious choosing.

These principles apply in every field of human activity. Unconsciously taking for granted that building is art, it is through working with volunteers that I have become consciously aware of it, and of the implications of following this path – of standing conventional values on their heads.

It seems more important than twiddling thumbs.